URANIUM GEOLOGY OF THE MIDDLE EAST AND NORTH AFRICA

URANIUM GEOLOGY OF THE MIDDLE EAST AND NORTH AFRICA

Resources, Exploration and Development Program

FARES HOWARI
Professor of Environmental Sciences and Engineering and Dean of College, Natural and Health Sciences at Zayed University, Dubai, United Arab Emirates

ABDELATY SALMAN
Nuclear Materials Authority, Egypt

PHILIP C. GOODELL
Professor of Uranium Geology and Director of Center for Entrepreneurial Geosciences, The University of Texas at El Paso, Texas, United States

ELSEVIER

Elsevier
Radarweg 29, PO Box 211, 1000 AE Amsterdam, Netherlands
The Boulevard, Langford Lane, Kidlington, Oxford OX5 1GB, United Kingdom
50 Hampshire Street, 5th Floor, Cambridge, MA 02139, United States

Notices
Knowledge and best practice in this field are constantly changing. As new research and experience broaden our understanding, changes in research methods, professional practices, or medical treatment may become necessary.

Practitioners and researchers must always rely on their own experience and knowledge in evaluating and using any information, methods, compounds, or experiments described herein. In using such information or methods they should be mindful of their own safety and the safety of others, including parties for whom they have a professional responsibility.

To the fullest extent of the law, neither the Publisher nor the authors, contributors, or editors, assume any liability for any injury and/or damage to persons or property as a matter of products liability, negligence or otherwise, or from any use or operation of any methods, products, instructions, or ideas contained in the material herein.

British Library Cataloguing-in-Publication Data
A catalogue record for this book is available from the British Library

Library of Congress Cataloging-in-Publication Data
A catalog record for this book is available from the Library of Congress

ISBN: 978-0-323-90992-1

Publisher: Candice Janco
Acquisitions Editor: Amy Shapiro
Editorial Project Manager: Sara Valentino
Production Project Manager: Bharatwaj Varatharajan
Cover Designer: Matthew Limbert

Typeset by Aptara, New Delhi, India

Contents

About the authors

Fares Howari, Professor and Dean of College, Natural and Health Sciences at Zayed University in United Arab Emirates.

Dr. Howari is an environmental scientist and engineer with a wide range of expertise. During the progressive years of his career, he developed distinguished academic, and project coordination skills. He has valuable experience in natural resources management and exploration, strategic planning and administration, and also in creative leadership. He currently serves as a Dean of College of Natural and Health Sciences at Zayed University. He served as Professor and Chair of the Department of Applied Sciences and Mathematics at Abu Dhabi University, United Arab Emirates, and a Director of Abu Dhabi University Center of Excellence of Environment, Health, and Safety. Dr. Howari also served at the University of Texas, Permian Basin, Center for International Energy and Environmental Policy, and the Bureau of Economic Geology at the University of Texas at Austin, Austin, Texas. Prof. Howari worked also as an Associate Professor at United Arab Emirates University. Prior to this, Dr. Howari worked at Texas A and M University, Texas, where he worked as a research scientist on the development and application of computer models and remote sensing techniques to assess environmental contamination problems.

Abdelaty Salman, Emeritus Professor, Nuclear Materials Authority, Egypt.

He completed his Ph.D. in Geology (Field and Structural Geology and Radioactivity) in 1975, Faculty of Science, Ain Shams University, Cairo, Egypt. Before joining the Nuclear Materials Authority, he worked as Head of Production Sector (1997); Vice Chairman of Nuclear Materials Authority (1997–1999); Chairman of Nuclear Materials Authority, Cairo, Egypt (1999–2000); and as a Private Consultant Geologist. He has authored 40 published articles, 28 unpublished scientific reports, and 3 books covering various geological fields including uranium exploration geology, radiometric survey, evaluation of uranium deposits, mining geology, and environmental and engineering-geology site studies for engineering projects.

Philip C. Goodell, Professor and Director of Center for Entrepreneurial Geosciences, the University of Texas at El Paso, Texas, United States.

His expertise lies in the Entrepreneurial Geosciences, a field of endeavor with unique properties. After obtaining Ph.D. from Harvard University, Cambridge, Massachusetts, he worked for many institutions in the United States of America. He founded the Center for Entrepreneurial Geosciences at the University of Texas at El Paso to provide opportunities for promising students with interests and capabilities in entrepreneurial geosciences. He is well known for his contributions in geochemistry, economic geology, mineralogy, semimetal and sulfosalt geochemistry, and uranium in volcanic and volcaniclastic rocks. The themes of Dr. Goodell's research in mineral deposit include: (1) Geochemical zoning in epithermal deposits (metal ratios), (2) volcanogenic uranium deposits and a model of their formation, (3) metallogeny of northern Mexico, (4) mega data management such as NURE, (5) remote sensing, (6) the use of multiple sensors simultaneously for detailed mapping in open-pit mines, and (7) multivariant analysis and large data bases.

CHAPTER 1

Introduction

CHAPTER 1.1

Uranium mineralogys

This part highlights general aspects of the geochemistry and mineralogy of uranium. It includes description for its geochemical behavior and the mode of distribution in some geological environments in the earth's crust. The primary and secondary uranium minerals are also discussed. Uranium is a dense, heavy, metallic element that naturally occurs within soils, rocks, and water in more than 200 mineral forms. In the earth's crust, uranium is of approximately 2.5 part per million (ppm), while in sea water is about 0.003 ppm. It exists in sea water at very small but variable levels, which ranges from less than 1 ppm to about 8 ppm.

In a variety of common rock-forming minerals, uranium is found to be in trace amount and is also present in minerals such as pitchblende (uranium oxide), coffinite (uranium silicate), and uraninite which represent the famous primary uranium minerals. Natural uranium is almost as heavy as gold with a specific gravity of about 18.7.The established geochemical and physical properties of natural uranium are strong predictors of the form in which the element is distributed in rock-forming minerals of the earth's crust and in different geological environments. Uranium is naturally radioactive, which means it decays spontaneously to form the steady element, lead, into a variety of the so-called "daughter elements." Throughout the decay, uranium and its daughter elements emit typical gamma and alpha radiation patterns that can be calculated even at extremely low levels and used to map the background concentrations levels in rocks and soil.

Uranium occurs in nature in many isotopic forms, which means that the uranium atomic nucleus contains 92 protons and together the atomic mass is augmented by a larger number of neutrons. U-235, U-238, and U-234 are the primary isotopes. Natural uranium's weight contains mainly U-238 atoms (99.3%), while U-235 (0.7%) forms the remaining weight, and U-234 (0.006%) is quite low in amount. Types of uranium deposit

Uranium Geology of the Middle East and North Africa
DOI: https://doi.org/10.1016/B978-0-323-90992-1.00001-4

have evolved considerably from the Archean to the Present. The key global drivers were as follows:

- The geotectonic transition during the Late Archean period.
- Strong rise in oxygen in the atmosphere from 2.4 to 2.2 Ga (giga annum or billions of years).
- Creation of Silurian land-plants. Other major differences in the types of uranium deposit are due to particular conjunctions of conditions such as those in the Cretaceous during sedimentation of phosphate. The mechanism for the fractionation of uranium on the earth had developed over several major periods.

The first one, between 4.55 and 3.2 Ga, refers to the formation of a thin, basically mafic crust in which trondhjeimite-tonalite-granodiorite the most fractionated rocks exceeded uranium concentrations of at least a few parts per million. In addition, uranium is basically found in refractory minerals and free oxygen is absent, during this time no uranium deposit could be assumed to have formed.

The second phase is distinguished by many widespread pulses of highly fractionated potassic granite, heavily enriched in Th, U, and K, from around 3.1 to 2.2 Ga. Peraluminous granite was selectively enriched in uranium late in this time, and to a lesser degree in potassium. They were the first magmas of pegmatite and granite capable of crystallizing uraninite at high temperature. Thorium-rich uraninite was released from these granitic suites, which would then be contained within large continental reservoirs along with pyrites and other heavy minerals (e.g., monazite, zircon, oxides of Fe-Ti). The absence of free oxygen prevented oxidation of the uraninite which, though only during these times, formed the oldest type of economic uranium deposit on the planet.

Records of the third period raised the amount of oxygen from 2.2 to 0.45 Ga to almost the current atmospheric level. Uraninite tetravalent uranium has been oxidized to hexavalent uranium that produced highly soluble uranyl ions in water. Uranium has been extensively trapped as a result of biological proliferation and huge amounts of organic matter accumulated in reduced epicontinental sedimentary successions, particularly during the Late Paleoproterozoic period.

A series of deposits of uranium were produced by redox processes, the first of which developed in Oklo area of Gabon, at a formational redox boundary at approximately 2.0 Ga. All known deposits of sodium metasomatism–related economically important uranium are around 1.8 Ga in age. During the Late Paleoproterozoic to early Mesoproterozoic, the high-grade, large tonnage unconformity-related deposits were also formed essentially.

The last period (0.45 Ga–Present) was coincidental to continental plants colonization. Intra-formational reduced traps representing the detrital accumulation of plants inside continental siliciclastic layers, in another uranium deposit family, basically only developed during this period: roll front, basal, tabular, and tectono-lithological types. Nonetheless, increased identification during diagenesis of hydrocarbon and hydrogen sulfide migration from oil or gas reservoirs indicates that sandstone-hosted uranium may be present in permeable sandstone that is older than the Silurian sandstone. Only during this last time are large deposits of uranium associated with high-level hydrothermal fluid and the deposits associated with evapotranspiration known due to their deposition in near-surface conditions, likely with a low potential for preservation.

Geochemistry of uranium

The production of concentrations of uranium that can be exploited depends on several geological variables, and also on climate in some areas. Involved geochemical processes are primarily actuated by orogenic activities. Therefore, a study of uranium geochemical activity and the tectonic settings that appear to be optimal for uranium enrichment is important. Data concerning uranium abundance and distribution are summarized as follows to provide a basis for discussion.

Distribution of uranium

Statistics on rocks and fluids with respect to uranium content are fairly scarce, and therefore it is calculated that the concentration and distribution of this element within the earth's crust are approximations, though possibly of the appropriate magnitude order. It is estimated that the concentration of uranium in the earth's crust is 0.0002%. Table 1.1.1 shows the uranium content in different rocks of the earth's crust.

Table 1.1.1 Concentration of uranium in the rocks of the earth's crust.

Rock type	Uranium concentration (ppm)
Ultra basic rocks	0.001
Basic rocks	0.5
Syenitic rocks	3
Granite rocks	3.5
Black shale	3.2
Sandstones	0.45
Limestone	2.2
Oil	0.1
Sea water	0.003

The uranium element is especially instructive in this respect for at least five factors that are as follows:

1. Uranium is an element in a trace quantity (approximate 2–3 ppm in the earth's crust; Table 1.1.1), which shows way of migrating and concentrating the elements.
2. Uranium occurs in two specific valence stages of opposing behavior, 4^+ and 6^+, also in the less frequent 5^+ state, and thereby emphasizing the importance of near-surface redox gradients in the diversification of minerals.
3. Uranium is an element of significant technical and environmental significance; therefore thermochemical parameters, physical properties, and phase equilibria of uranium minerals, aqueous species were studied in depth.
4. Uranium mineralogy is altered indirectly and directly by the microbial and other biological activity, and thus highlights the Earth's geo- and biospheric mineralogical co-evolution and
5. The neutronic characteristics of 235U are such that naturally occurring fission and related neutron-capture reactions will result in addition of trans-uranium elements to crustal abundances (e.g., approximately 3 metric tons of Pu at the Oklo's natural reactors, 5 metric tons of Pu from atmospheric testing of nuclear weapons, and 1800 metric tons of Pu from nuclear power reactors). Furthermore, the progressive transition of uranium minerals from uranium to radiogenic Pb leads to a range of fascinating and idiosyncratic actions of alteration.

The story of mineral evolution of thorium offers an instructive addition to uranium because of its relative insolubility 4^+ valence states, in the crystal chemical industry of these elements, and thorium does not adopt the extremely soluble 6^+ valence state of uranium. The story of thorium is quite similar. In much more limited circumstances, Th^{4+} (aq) is mobilized than U^{4+} (aq). Aqueous solution reveals a substantial difference under complexity. Both thorium and uranium are special and essential to geoscientists in the periodic table. Thorium and uranium are the two most massive natural elements (without counting the very small concentration of contemporary crustal concentrations of certain other heavy elements, e.g., Pu and Pa), and the most abundant in the actinide series. The earth's radioactive decay has caused the thermal evolution of uranium and thorium, contributing to the layered internal structure of the earth and the continuing movement of plates across the earth's surface.

The three naturally occurring radioactive decay series, the 238U, 235U, and 232Th, also offer a wide variety of radiogenic elements, such as radon. Pb's decay into stable isotopes forms the basis for geochronological measurements relating absolute dates to the geological time scale at the start of twentieth century. The areas of fission-track data rely on the spontaneous fission nuclear processes and He's creation by alpha decay. A wide variety of geological process can lead to departing from secular equilibrium in the decay chains and provide the basis for the geochemical field of uranium series. The main element for nuclear power is uranium and thorium. 235U is fissile, 238U can be used to "breed" 239Pu which is fissile, and 232Th to "breed" 233U which is also fissile. Over 220 different minerals of uranium have been identified, but uranium mineralogy is dominated by uraninite (ideally UO_2) in a volumetric form. The past of uranium mineralogy, including its distribution and diversification, is also nearly related to deposits of ore. Observations of the ore deposits in this study are a key to understand mineral evolution of uranium and thorium mineral. The origin of deposits of uranium ore on the earth can be broken down into four main phases.

The concentration of both uranium and thorium was elevated in zirconium (which can concentrate uranium compared to thorium) into the most enriched magmas fluids in the early stages, the Hadean and previous Archean eons (~4.5–3.5 Ga). Following the enrichment of the liquid the first mineral uranium, mainly uraninite and coffinite, was formed (ideally $USiO_4$).

In the second stage, the thorium-bearing detrital uraninite was formed in the quartz-pebble conglomerates type Witwatersrand, which were deposited in an anoxic soil from about 3.5–2.2 Ga. Abiotic altered coffinite and uraninite that may result in a small number of uranyl oxide-hydroxides and other uranyl forms, including a major "self-oxidation" mechanism by which radioactive decay contributes to the formation of daughter products such as Pb and the production of He from alpha particles.

The third step of the earth mineral evolution of uranium preceded at ~2.2 Ga by the Great Oxidation Event (GOE) and was subsequently triggered by biological processes indirectly. During the process, low-thorium uraninite was deposited from a saline and hydrothermal (100–300 °C) oxidizing solution, which transported uranium to 6^+ oxidation state as compounds $(UO_2)^{2+}$. Sources include deposited of nonconformity- and vein-type uranium (Canada and Australia) and the special Oklo natural nuclear reactors in Gabon.

Over this time, the majority of uranyl minerals may form through weathering processes in the near-surface O_2-bearing setting. The following changes were strongly influenced by microbial activity in near-surface uranium mineralogy, both indirectly due to changes in atmospheric chemical processes, and directly by passive sequestration of uranium and uranium-coupled redox metabolism. The increase in land plants approximately 400 million years ago resulted in the fourth stage of the deposition of uranium mineral, as low-temperature, oxygenated, uranium-rich near-surface waters have precipitated uraninite and coffinite in organically rich continental sediments. The resulting ore deposits of sandstone type represent a significant fraction of known uranium stocks of the United States America and Australia. The uranium reserves of the western United States America are primarily roll front stockpiles; whose distinctive curved form is dominated by the movement of groundwater by coarse-grained sandstones. The current "technogenic" nuclear evolution of uranium and thorium represents the fifth and final step, but is not addressed in depth in this regard. The production of nuclear power generation reactors (providing approximate 15% of global electricity) depends on the use of fissile 235U.

Mineralogy of uranium and thorium

M. H. Klaproth (1743–1817), a German chemist and mineralogist discovered the element uranium in 1789. The leading chemical researcher at his time was Klaproth, the first professor of chemistry at the University of Berlin. A pitchblende yellow oxide (UO_3) from the mine at Johanngeorgenstadt in the Erzgebirge (Saxony) was prepared by Georg Wagsfort. Pitchblende was previously recognized as a separate mineral group, pechblende from this location and nearby town of Joachimsthal, but the composition of its elements, variously considered to include zinc, iron, and tungsten, was not known. Klaproth heated carbon with this yellow oxide and found it to be reduced to an ingredient when he obtained a dark powder, somewhat metallic. This powder, which he called uranium, was added to the natural ore by the name uraninite. Klaproth's product has been thought to be the element for approximately 50 years, till 1841, before the French chemist E. Peligot demonstrated that it was actually the oxide, UO_2. Through reducing $UC1_4$ with potassium, Peligot prepared the first elemental uranium.

Uranium mineralogy

Uranium is known as a lithophile element and has twice the average crustal abundance in granitic rocks. Nevertheless, granite weathering is the largest uranium source in sedimentary rocks at present. Although valence states are between 2^{+} and 6^{+}, only the 4^{+} and 6^{+} valence states are crucial in minerals. There is some proof of a 5^{+} state as both a solution species and in some of the uranium oxides, but its position is not proven in mineral structures. Minerals of uranium are generally classified in two major classes—known as "primary" and "secondary" minerals. This distinction relies on the original assumption that all uranium is first deposited as uraninite, which was the only truly economic mineral, and then oxidation resulted in other minerals being produced. Many uranyl minerals are now accepted by sources without primary uraninite and created adequate concentrations for economic use. There have also been reports of some new minerals of uranium that have proven to be ore minerals in some big deposits.

Mineralogically, the division of the uranium minerals is probably best based on the valence condition of uranium. Therefore, a family of U^{4+} is "reduced" in which all or most of uranium occurs as U^{6+} and family of "oxidized" is the dominant valence state. The bulk of U^{6+} minerals contains the $UO{\sim}^{+}$ uranyl ion. By using anionic unit or related groups, the oxidized unit structure separates the mineral chemically. To describe the minerals initially deposited and the alteration minerals, the term "primary" and "secondary" are to be used. The bulk of uranium minerals are present in various ore deposits. Usually only two reduced minerals exist at a given deposit. The oxidized minerals present in the deposit are subject to Eh-pH conditions and reactive anions are available. In the absence of reactive anions, hydrated oxides and uranates form.

However, uranyl ion is relatively soluble, and groundwater can effectively spread it from the reduced source. The uranyl minerals deposited are complex compounds of exiting oxyanions. The formation rate of minerals deposit on mine walls can be extremely rapid, and could form within several months. Mineralogy is usually zoning in all deposits, which is followed by complex uranates and, ultimately, uranyl oxysalts, with reduced mineral hydrated oxides within the first few centimeters of the oxidation zone. The particular oxysalt relies on the availability and solubility of the given anion, the most common being silicates and phosphates and the comparatively uncommon carbonates and sulfates relatively rare.

Any of these minerals may have uranium (more than 1% uranium) as a key element. Uranium is tetravalent for some of these minerals and is hexavalent for the most part. Simple oxides and hydrated oxides, complex titanates and niobates, silicates, phosphates, carbonates, sulfates, vanadates, and arsenates are such uranium minerals. Unidentified uranium compounds can occur in some marine carbonaceous shale, lignite, and coal and perhaps as inter-granular films in igneous rocks. The main uranium minerals in the larger ore deposits are oxides pitchblende and silicate coffinite, uraninite, vanadates carnotite and tyuyamunite, and complex titanates brannerite and davidite. The original element or simple naturally occurring sulfide, arsenide, or telluride is not classified as uranium.

Mineral evolution concepts are elucidated by the origins and near-surface distributions of known thorium and/or uranium minerals. The history can be broken down into four parts. The first, from approximate 4.5 to 3.5 Ga, contained amounts of uranium and thorium from their initial uniform distribution of trace magmatic fluids from which the first U^{4+} and Th^{4+} minerals, thorianite (ThO_2), uranine (ideally UO_2), and coffinite ($USiO_4$) were precipitated into the crust.

The Witwatersrand-type quartz-pebble conglomerates deposition in a highly anoxic fluvial setting produced a significant low-grade concentration of detrital uraninite during the second phase, from about ~3.5 to 2.2 Ga. Coffinite and uraninite abiotic alteration may have led to the formation of a small collection of uranyl oxide-hydroxides, including radiolysis and auto-oxidation generated by radioactive decay and helium production from alpha particles. The Great Oxidation Event (GOE) was about 2.2 Ga and was thus indirectly induced by biological activity. In the third stage of mineral evolution of uranium most common uranium minerals were precipitated by soluble uranyl $(U^{6+}O_2)^{2+}$ complexes. The majority of uraninite deposited during this point is low in thorium and precipitated with the transport of $(UO_2)^{2+}$-chloride complexes in saline and oxidizing hydrothermal solutions (100–300 °C). Examples include the uranium deposits (Canada and Australia) of unconformity- and vein-type and Gabon's special Oklo natural nuclear reactors. The start of $(UO_2)^{2+}$ complex hydrothermal transports in the upper crust indicates the accessibility of $CaSO_4$-bearing evaporites following the GOE. During this step, the majority of uranyl minerals may have for the first time formed through weathering processes in the O_2-bearing near-surface climate.

Uranium mineralization fourth stage started around 400 million years ago, when the growth of the land plants resulted in organic-rich non-marine sediments supporting new deposits of sandstone ore. Accumulation modes, and also uraninite compositions, along with several uranium oxidation states (4^+, 5^+, and 6^+), are a sensitive predictor of global redox conditions. The behavior, in the absence of aqueous F-complexes of thorium, that has one oxidation state (4^+) with very low solubility, does not indicate a changing redox state.

Therefore, the thorium geochemical concentration relative to uranium is restricted to special, magmatically associated conditions, in which U^{4+} is ideally extracted by carbonate or chloride group, and mineral surface reactions at low temperatures. Uranium and thorium near-surface mineralogy measured the geotectonic and geobiological history of a planet. Thorium and uranium are not abundant enough to create its own minerals or ore deposits without extensive reworking of the upper mantle and crust magmatic-related fluid. However, all but some known minerals of uranium are unlikely to evolve in the absence of surface oxidation.

CHAPTER 1.2

Nuclear program in MENA

Many countries in the Middle East and North Africa (MENA) have programs and activities in the nuclear fields, but they differ from one another in both form and content. Most of them do not exceed the front stages of the nuclear fuel cycle, or utilize some peaceful applications of radioisotopes in medicine, industry, agriculture, animal production and irradiation of some foodstuffs to increase the period of resistance to damage.

Uranium ores, which are the main source of uranium needed for the nuclear fuel cycle, have not been proven to exist in many of these countries in a way that can be exploited economically. They often fall under the type of potential or predictive ores (i.e., they are very rough estimates). Therefore, the authors believe that each of these countries must have a clearly defined nuclear program—for peaceful applications—with a focus on the exploration and evaluation of uranium ores. It is possible for any integrated nuclear program to start without the presence of uranium ores, and there are also great difficulties in obtaining uranium in the producing countries.

Several countries in the MENA announced their plans for their future energy mix to use nuclear power. In developing the first Arab nuclear power station, the United Arab Emirates (UAE) is ahead of its region, becoming the first country in 27 years to build its first nuclear reactor. These countries are forced to consider the highly contested cost and advantage of nuclear energy when trying to meet their growing energy needs.

New nuclear power plants are scheduled to be operational throughout the MENA area within the next decade. The UAE's Barakah flagship project will have a total installed capacity of 5.6 gigawatts (GW); the first unit started producing electricity in 2020 delivering almost a quarter of the nation's energy needs. In Saudi Arabia, 16 nuclear reactors with the combined capacity of more the 17 GW (expected to fulfill 15% of the country's energy needs) are going to be initiated with the most ambitious nuclear program by 2032. In 2022, the first reactor is expected to be functional. The first nuclear power plant in Jordan with a 2000 megawatts (MW) range is expected to be operational in 2023, signed by Jordan with Russia's Rosatom, the state nuclear corporation. Egypt has also signed an agreement with Rosatom to build four reactors, each with a capacity of 1200 MW, in the course of the next 12 years. This is the nominated El Dabaa nuclear power plant project. Tunisia, Morocco, and Algeria, which have been running two research reactors, Draria and Ain Oussera, since the early 1990s, are also evaluating their options. However, after the Fukushima Daiichi Nuclear Power plant accident in Japan in 2011, Oman, Kuwait, and Qatar cancelled their nuclear plans.

For more than 50 years, nuclear power has failed worldwide to retain its growth momentum. Its popularity has often been dampened for significant periods by major accidents. Despite technological advancements since the first peaceful use of nuclear power ended in the 1950s, many challenges remain, from cost to safety. Every argument in favor of nuclear power has an equally persuasive argument against it. The nuclear debate has therefore taken place in cycles: from nuclear power supporters to nuclear power opponents who recover and recapture nuclear power, limiting the contribution of nuclear power to world energy. In terms of nuclear power, the MENA area faces unique challenges and opportunities. The volatile geopolitical environment makes nuclear technology much more contentious in the region: countries fear that neighbors will take military action on their civil nuclear programs. For example, Iran's nuclear plan led to sanctions being imposed by the world community and years of negotiations. With the already heavily subsidized economies in the country, MENA governments are also worried about these large nuclear investments. In comparison to most other nuclear energy users worldwide, some countries in the Middle

East are rich in fuel and gas, allowing them more time to consider their role in meeting the energy needs of their nuclear power. Nuclear power will enter the MENA region energy mix but will not resolve the long-running energy problems in the region without much-needed economic reforms. Quite bluntly, civil nuclear technology is not an escape path. It can help, but the reality that MENA countries must face is that modern, competitive economies must be constructed. Nuclear power could well be more of a burden than a benefit without this necessary transformation.

The role of nuclear programs

Role of nuclear programs in providing water

The water problem is one of the most important issues, and the countries of the MENA prioritize it accordingly, as its short-term shortage and scarcity in the long run threaten development projects in the MENA countries. Reviewing studies on population of the Arab world included in the MENA region indicate that these countries extend from the Persian Gulf in the east to the Atlantic Ocean in the west, with a total population of approximately 400 million in January 2016. The population growth rate in that region ranges between 2.4% (Lebanon) and 3.3% (Algeria). The largest country in terms of population is Egypt, with a population growth rate of 2.7% and a population of 99,375,741 people, followed by Iraq and Saudi Arabia. The economic growth rate in these countries does not fully support the various requirements of population growth especially in terms of their energy and water needs. In the MENA countries, we find that the water uses in agriculture occupy the highest percentage, followed by daily and industrial uses.

The importance of the water issue in these countries is well known as the situation facing the countries of the MENA is very critical, for example, there are 11 of the 20 Arab countries in this region that currently use more than half of their water sources. Libya and the countries of the Arabian Peninsula except Oman use 100% of its water sources, and they depend on expensive seawater desalination or groundwater withdrawal with the expectation of increasing the population in an alarming way. To find out how serious the issue will be in the future, we will review what the situation will be in 2025 based on recent studies of the population census and estimates of renewable water sources. For example, we find that Lebanon will face pressure with respect to water, as it will reach the share of every 600–1000 individuals 1 million cubic meters.

As for Morocco, Egypt and Syria, water scarcity will occur, as the share of every 1000–2000 people will reach one million cubic meters of water. As for Bahrain, Algeria, Djibouti, Kuwait, Jordan, Libya, Qatar, Oman, Saudi Arabia, Tunisia, Somalia, the United Arab Emirates, and Yemen, they will meet a water barrier before 2025, when one million cubic meters of water is distributed to more than 2000 individuals. It is clear from the foregoing that there is a grave problem with regard to the water shortage expected to occur in the MENA countries during the next 20 years and beyond, and therefore it must—from now—work to avoid this catastrophe by working to establish nuclear plants to desalinate sea water to increase the water sources in these countries.

The role of nuclear programs in providing energy

Energy represents the most critical challenge of the twenty-first century as it is considered the main life requirement and is indispensable in daily life or industrial and development projects, which translate in the end to raising the individual's standard of living and welfare. There are three main sources of energy: (1) energy from fossil fuels (non-renewable) such as petroleum, natural gas, and coal; (2) new and renewable energy, which includes hydro, solar, wind and geothermal; and (3) nuclear sources. Table 1.2.1 shows the global energy consumption during the period from 1980 to 2006, indicating a continuous increase in energy consumption at the Global level.

As for the new and renewable energy, its sources include hydro, sun, wind, geothermal and bioenergy. But these sources contribute a modest share as sources of energy because many countries have already exploited their water resources to generate energy, and only a small amount remains, and these sources are considered cheap and environmentally clean at the

Table 1.2.1 Global consumption of primary energy sources.

Average power (terawatt)			
1980	**2004**	**2006**	**Source type**
4.38	5.58	5.74	Petroleum
1.8	3.45	3.61	Gas
2.34	3.87	4.27	coal
0.599	0.933	0.995	Electric energy
0.253	0.914	0.929	Nuclear energy
0.016	0.133	0.158	Renewable energy
9.48	15	15.8	Sum

same time. As for other new and renewable sources, they are still in their first generation, and they need many studies to reduce the costs of obtaining them, and they can add a "limited" part to the total required energy. The third source of energy—which is considered by the author to be an important source—represents nuclear energy. This source can replace the equivalent of hundreds of millions of tons of oil equivalent.

The global energy demand by 2040 will increase by 30% from what it is in 2017. This is an expectation based on two main influences, which are the growth of the global economy at a rate of 4.3% per year, and an increase in the world population from 7.8 billion to more than 9 billion in 2040. In the sustainable development scenario, the share of energy sources with low carbon emissions in the energy mix doubles to 40% by 2040. The demand for coal will start to descend, and the demand for oil will reach its peak a little after that. And as electricity generation becomes less dependent on carbon sources by 2040, it will depend increasingly on renewables (with more than 60%) and nuclear energy (15%), as well as on carbon removal and storage (6%). There are several considerations for concern regarding nuclear energy in the MENA region.

In social terms, these countries need many large urban and industrial projects to raise the standard of living of the individuals in these countries. These development projects require enormous energy not only to build them but to operate them and ensure their maintenance. Egypt is an example of the implementation of many giant development projects in various Egyptian regions, especially mining and industrial investment to serve the Egyptian community. The primary application of the use of nuclear energy for peaceful purposes is the production of electric energy and desalination of sea water in the MENA region. The region relies on oil, minerals, and agriculture as a major source of national income; construction projects require an enormous amount of energy and water to establish, operate, and sustain them. Therefore, dependence on nuclear energy for use and operation is justified in the field and socially.

In economic terms, the high costs of establishing nuclear plants represent the main obstacle to nuclear energy, but its advantage is the amount of energy inherent in nuclear fuel, which makes the cost of producing kilowatt hours very low in comparison to fossil fuels. For example, a total consumption of an amount equivalent to a few grams of uranium is enough to light 20,000 lamps for 12 consecutive days. Therefore, relying on nuclear energy to meet electrical energy needs has a reasonable economic return, despite the availability of large quantities of oil. But it is worth noting that oil prices are variable and sometimes rise dramatically, as they often depend on the prevailing political conditions in the world. It is a non-renewable energy source and MENA countries should not remain dependent on oil as a primary source of energy until it is depleted. If there is no ready substitute, it would be an energy disaster.

As for investment considerations, it is known that establishing nuclear plants, providing nuclear fuel, treating and manufacturing it, and disposing of radioactive waste requires huge costs, therefore it is necessary to provide huge funds to invest in this nuclear industry. In the opinion of the author, the costs needed to establish these stations can be managed in cooperation between some countries in the region that have financing possibilities based on the common interest. There are many MENA countries that have the necessary financing and can invest what they pay on an economic basis, in addition to raising the technical level, which will be reflected in both the financing countries and the host countries.

From an environmental point of view, it is known that nuclear power reactors do not emit harmful gases in significant quantities, such as those emitted from fossil fuel stations, the most dangerous of which is carbon dioxide, which significantly affects climate changes and causes global warming. To illustrate this issue, the average increase in the earth's temperature has increased by between 1 and 5 °F since the late nineteenth century. In the twentieth century, it became evident that there were 10 warm years from the last 15 years of the century, and that 1998 was the warmest in terms of climate records. Likewise, the ice coverage in the northern hemisphere and floating snow in the Arctic Ocean have decreased. At the level of the globe, the level of the surface of sea water has risen between 4 and 8 inches above its levels in the last century. Also, rainwater deposits worldwide increased by 1%, and their precipitation rates in a severe manner have increased. Therefore, it can be emphasized that nuclear power plants have the ability to compensate for the imbalance resulting from the emission of harmful gases into the environment to restore the environmental balance according to the levels required during the period from 2008 to 2012 according to the Kyoto Protocol and to reduce emissions of harmful gases to the environment. It should be noted that preserving the environment is not a kind of luxury—as some think—but it is a priority first for humanity and second for the surrounding environment. The MENA countries have the material and human capabilities to establish nuclear power plants with steady and effective steps.

As for political and strategic considerations, the introduction of nuclear energy does not necessarily imply its use in non-peaceful areas. It is extremely important that these countries significantly develop their nuclear energy capabilities, and promote its peaceful use to meet their electrical energy and desalination needs, as well as support other vital and important projects necessary for the development of society and the environment. Time is an "important" and "influential" element.

Nuclear reactors

A nuclear reactor is a metal container in which there is what is called the core of the reactor consisting of nuclear fuel and control rods in addition to other reactor components. Many reactors are known as thermal reactors, because these reactors rely on thermal neutrons to induce fission, and this is done by providing an opportunity to collide uranium-235 nuclei when released with other light matter nuclei called a sedative, usually using regular water, heavy water, or Graphite as a sedative.

The thermal energy produced by the fuel is used to generate the steam needed to make electricity. This is achieved by the movement of a liquid or gas in the core of the reactor called the coolant. The coolant absorbs the heat created in the nuclear fuel, thereby raising its temperature and then travels through pipes within the so-called "steam generator" used for many types of the stations. The coolant heat travels to the water surrounding the pipes, which contributes to the production of steam that is used to control turbines, which in turn handle the electric generator. The coolant then returns after losing a large part of its heat during the heat exchange process, generating the steam to the core of the reactor and the cycle is repeated. This fuel cooler has a primary cooling cycle. The steam generation system used to fuel the turbines, which in turn drive the generator, dictates the efficiency of the nuclear plant. This steam can be produced directly within the reactor core, as in the case of boiling water reactors, or produced by a medium called a steam generator, as in the case of pressurized water reactors, as follows.

Pressurized ordinary water reactors

This type of reactor is one of the most common reactors in the world for electricity generation. The reactor consists of the following components:

1. The core of the reactor in which the nuclear fuel is in the shape of rods, known as the nuclear fuel rods, as well as a number of control rods to control the chain reaction. The fuel used in most pressurized water reactors is produced from enriched uranium dioxide, 3%, and makes the

fuel, which contains a ratio of the isotope of uranium-235. It is in the form of small cylinders with a length of about 1 cm and a diameter of 0.8 cm. These cylinders are filled from long tubes known as fuel casings. The fuel casing is made usually from zirconium alloy or non-corrosion steel due to the ability of these materials to deliver heat and their high resistance to corrosion and rust, and because of their low susceptibility to neutron absorption. Several fuel columns are collected using a square metal mesh known as a fuel bundle, and this metal mesh allows passage of the coolant between the fuel bundles and contact with the fuel columns. The reactor capacity is related to the number of fuel bundles in it (Fig. 1.2.1).

2. The containment container contains the core of the reactor and the water that is used to calm the neutrons involved in the reaction and to get rid of the heat resulting from the fission process. It is covered from the inside. This container is made of strong steel, about 20-layer thick with a non-corrosive alloy.

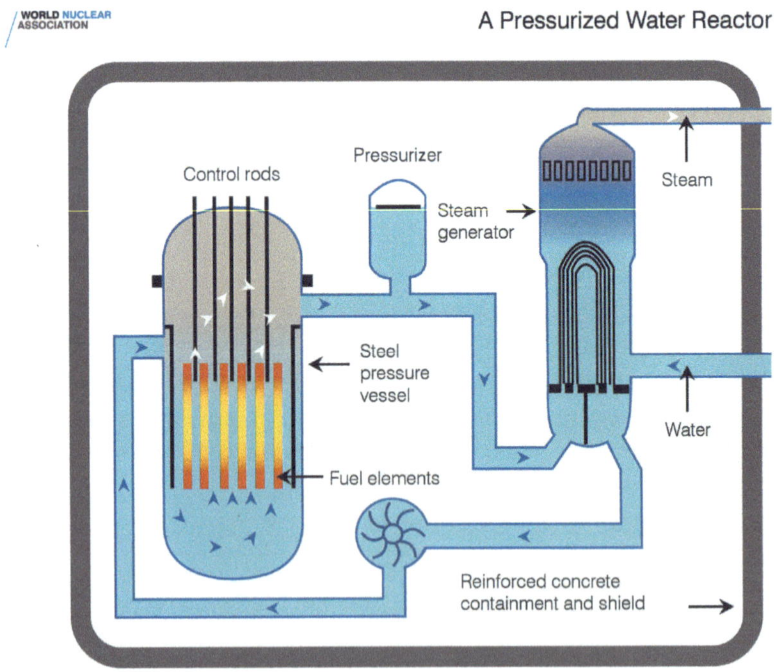

Fig. 1.2.1 ***Pressurized water nuclear power reactor.*** *(Courtesy: Dr. Jonathan Cobb, Senior Communication Manager, World Nuclear Association.)*

3. The cooling and energy transfer system inside the reactor core is based on the following three consecutive circuits:
   a. *The primary circuit:* It consists of 3 identical cooling circuits in the case of reactors of 900 MW, and 4 identical cooling circuits in the case of reactors with a capacity of 1300 MW. Each of them passes through the containment container, and the heat generated inside the core of the reactor is transferred to heat exchangers called steam generators by which the first thermal energy transfer is carried out by these closed circuits. There is a pressure controller that works to control the pressure of the circuit, and regular water is used as a coolant and moderator in this type of reactor. It is not allowed to boil inside the primary cooling cycle even though the temperature of its heat is about 300 °C, which necessitates keeping the coolant under a high pressure of about 160 atmospheric pressure.
   b. *The secondary circuit:* This circuit works to absorb the heat in the first circle by means of heat exchangers and transfer it in the form of steam in the direction of the turbines. This represents the second transfer of energy (a closed circuit that does not contain radioactive materials).
   c. *The third circuit for cooling:* It contains a condenser that works to restore the latent heat in the steam coming out of the turbines, and this condenser cools either by using cooling towers or with water in the event that a water source—such as a sea or a river—is provided nearby the nuclear plant. The condensed cooling water does not contain any radioactive materials.

Boiling water reactors

This type of reactors consists of the following two parts:

1. *Ordinary reactor:* The core of the reactor contains the fuel that allows the passage of the coolant, which is water passing through the fuel rods under high pressure (70 atmospheric pressure). Under this pressure, the water is heated and turned into steam inside the reactor, and the steam is isolated from the water and then dried by a separator. The steam and the vapor dryers in the upper part of the reactor reach the appropriate properties of the turbine operation (70 atmospheric pressure, 0.25% humidity), and the nuclear reaction is controlled by control bars in the form of cross-panels (cruciform). These panels are inserted from the bottom of the reactor and permeate the core of the reactor

that is controlled automatically or manually. The fuel is a bundle with a number of enriched uranium oxide rods coated with zirconium alloy. The rate of combustion of this fuel reaches 27,000 MW per day per ton uranium, and the fuel is recharged annually after the reactor stops working. This type of reactor is characterized by the presence of jet pumps inside which make the cooling pump.

2. *The cooling circuit:* It is a direct circuit for cooling, in which the steam created from the reactor is directed toward the steam turbine, generating the rotational force of the electric generator. It then passes through the condenser, and it turns into water that returns back to the reactor. These reactors are characterized by (1) not having to use a steam generator; and (2) the reactor operating at a relatively lower pressure than pressurized water reactors. However, compressed water reactors are characterized by the presence of a separation of the initial cooling cycle water exposed to radiation from the steam generation cycle heading to the turbine that is considered safer (Fig. 1.2.2).

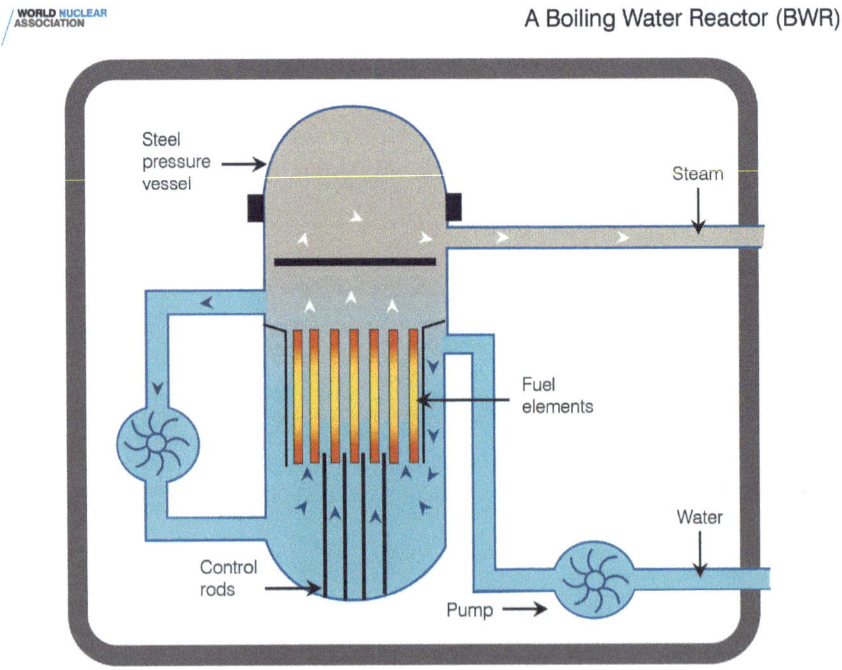

Fig. 1.2.2 ***Boiling water reactor.*** *(Courtesy: Dr. Jonathan Cobb, Senior Communication Manager, World Nuclear Association.)*

Water-cooled graphite reactors

Graphite is used in these types of reactors as a moderator for neutrons, and regular water is used as a coolant and the core of the reactor consists of clusters of graphite blocks inside tubes called pressure tubes. Inside these tubes, there are columns of nuclear fuel that passes around it, and cooling water that turns into steam that goes out to the turbines directly. The Chernobyl reactor, where the nuclear accident occurred on April 26, 1986 and resulted in a dangerous radiation leak, is this type of reactor. These reactors are only found in the Soviet Union (previously), and they differ from pressurized ordinary water reactors and boiled water reactors in not having a concrete containment container surrounding it and the components of the refrigeration cycle, so that it can bear pressure and temperatures in the event of emergency accidents, thus preventing radiation leakage to the environment.

Heavy water reactors

Heavy water (Fig. 1.2.3) is used as a sedative, and heavy water is a water type in which deuterium substitutes for hydrogen. Heavy water is distinguished

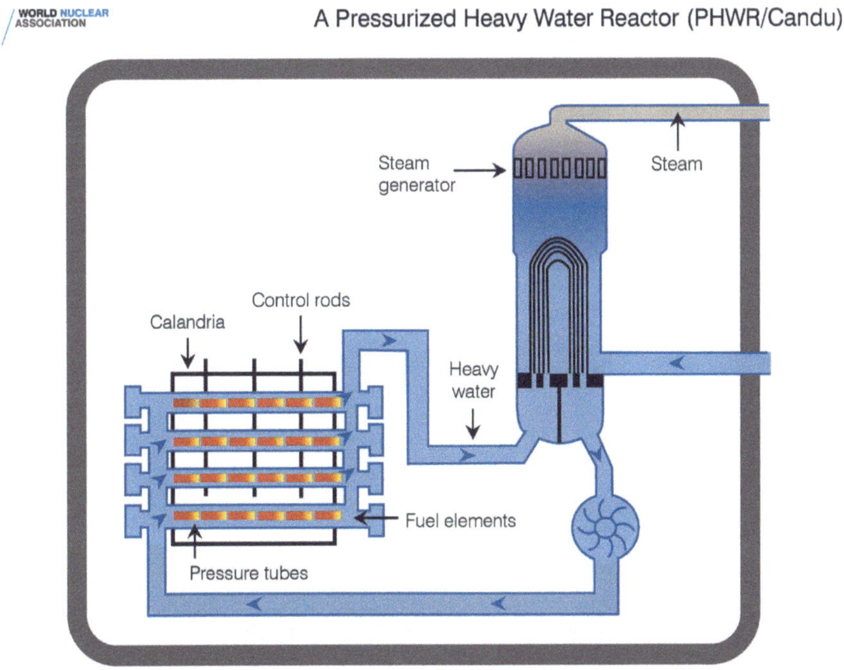

Fig. 1.2.3 ***Heavy water reactor.*** *(Courtesy: Dr. Jonathan Cobb, Senior Communication Manager, World Nuclear Association.)*

by a low neutron absorption value, which tends to maximize the number of neutrons available to induce fission, such that uranium can be used. The coolant flows over the fuel columns in hollow tubes called pressure tubes and is enclosed by calming water, keeping the coolant under high pressure inside these tubes without the need to use a pressure vessel. This reactor is characterized by the possibility of recharging the fuel without having to stop it.

Gas-cooling reactors

The use and development of gas-cooling reactors (Fig. 1.2.4) began in the United States of America during the World War II to convert uranium-238 to plutonium-239, which is used in military purposes. This type of reactor is considered the starting point of nuclear programs for many countries such as France and Britain.

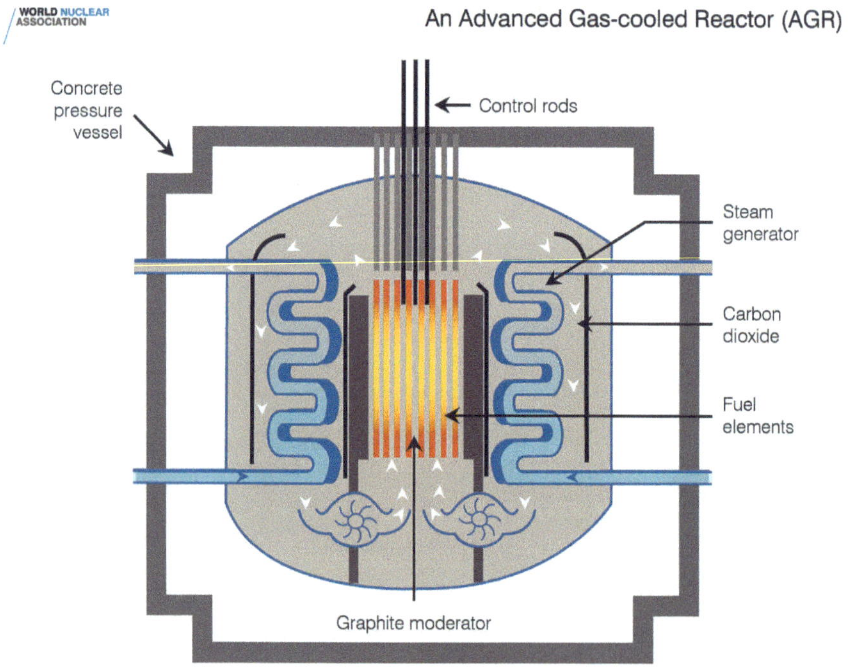

Fig. 1.2.4 ***An advanced gas cooling reactor (AGR): https://www.world-nuclear.org/information-library/nuclear-fuel-cycle/nuclear-power-reactors/nuclear-power-reactors.aspx.*** *(Courtesy: Dr. Jonathan Cobb, Senior Communication Manager, World Nuclear Association.)*

The following are among the most important types of gas-cooling reactors:

- Magnox reactors
- Advanced gas-cooling reactor

In the magnox reactors, natural uranium is used as a fuel, and it is made in the form of columns about 1 m in length and 2.5 cm in diameter. The name of this type of reactor is due to the use of magnox alloy formed of magnesium, aluminum, beryllium, and copper. It is characterized by its non-corrosion due to the effect of carbon dioxide used as a coolant. Graphite is used as a moderator, and it is in the form of blocks with channels in which the fuel columns enter through it and the control columns are made of steel with added boron. Their length is 5–10 m and the diameter is about 4–8 cm. As for the advanced gas-cooling reactors, enriched uranium oxide is used as fuel and delivered in the form of short hollow cylinders from the inside. These cylinders are packed in ribbed casings made of stainless steel added to the proportion of chromium and niobium. This type of reactor can be charged with fuel during operation without the need to stop it. As for high-temperature gas-cooling reactors, uranium containing a very high percentage of uranium-235 mixed with thorium-232 is used as fuel. The importance of using thorium is due to its conversion to uranium-233 isotope by absorption of neutrons; and because the uranium-233 isotope is fissionable, its formation in the reactor is an addition to the original fuel. It places the fuel in the form of short cylinders of uranium and thorium carbide, coated with a graphite used as a sedative, and uses helium gas as a coolant, which is characterized as an inert gas that does not interact with the moderator, no matter how high its temperature increases, because it does not absorb neutrons and thus does not turn into radioactive gas. It is worth noting that the establishment of this type of reactors is very costly and the fuel combustion rate used is low.

Rapid generating reactors

This type depends at the beginning of its operation on uranium as a fuel and then gradually replaces the plutonium formed in the reactor during operation until the final fuel of the reactor is only plutonium with natural uranium. The core of the reactor in this type of reactors is also surrounded by a casing of natural or depleted uranium (previously used in the operation of other reactors), and the importance of using natural or depleted uranium is due to its distinct ability to generate plutonium, thus providing additional fuel during operation and offsetting the original fuel combustion. Hence these reactors are named by the reactors generated, and as these reactors depend on power generation on nuclear fission by means of released neutrons at high speeds, they do not need to use a sedative like the aforementioned thermal reactors. One of the most important advantages of this type of reactors is that it uses fuel more efficiently, about 60 times more than other types, and therefore their utilization helps to rationalize the use of nuclear fuel and reduces the rate of global reserve consumption of uranium. The most important types of rapid reactors generated are rapid reactors cooled with molten metal, the fuel is a mixture of plutonium oxides and uranium made in the form of columns placed in stainless steel casings. Heat and the sodium are kept under normal pressure, because they do not boil at the high temperatures that the reactor core reaches (about 650 °C). To prevent the possibility of contact with radioactive sodium with water, it is passed to a second cooling circuit, usually cooled with sodium, which is directed to the steam generator to generate steam.

CHAPTER 1.3

Overview of uranium resources in the MENA countries

Using various related recognition criteria, the book reported the potential uranium provinces in the MENA countries. The uranium-rich provinces were identified based on geological and tectonic setting in conjunction with geochronological sequence and geochemical characteristics. This book will provide ideas and necessary data for the development of uranium exploration and production programs. One of the main outcomes of this research is to spotlight the promising potential uranium provinces in the MENA.

Special emphasis is given to Pan-African granites that are one of the most suitable environments to host vein-type uranium deposits. This case is very clear in Algeria, Morocco, Egypt, Sudan, Saudi Arabia, and Turkey. The uranium mineralizations are hosted in these granites within some favorable structures such as faults and fractures. The uranium minerals are mainly secondary (primarily uranophane) and occasionally some pitchblende and uraninites are present. They are associated with sulfide as pyrite, chalcopyrite, galena, sphalerite, and molybdenite. The gangues minerals are mainly iron, manganese oxides, and fluorite. The most common alterations associated with this gangue are hematitization, silicification with black silica veins, carbonetization, and the presence of fluorite. Volcanic rocks represented another potential source for uranium in many of the studied countries. The volcanics range in age from 74 to 302 million years ago and some of them host uranium or uranium/thorium mineralization.

The presence of intra-cratonic basins is also observed within many basement rock exposures in the study area. Often these basins are filled with late-Proterozoic molasses-type sediments such as the Hammamat series in Egypt and can form important uranium traps according to their geochemical and geological characteristics. This feature is well known in many of the studied countries especially in Algeria, Egypt, and Saudi Arabia. This book also documents evidence of the presence of intra-cratonic basins filled with Paleozoic sequences; the lower horizons of these sequences have potential for hosting uranium resources. This was noted in Libya (Morzok basin) and in Egypt (Wadi El Kharite basin). The unconformity uranium deposits could exist in some countries such as Algeria and Egypt. Good potential was found in Paleozoic rock sequences in many of the studied countries. In addition, surfacial uranium deposits were reported in Jordan nearby some uranium-bearing phosphate deposits. Phosphorite belts in North Africa and the Middle East can form additional non-conventional uranium resources in several geographic regions. Black shales nearby the phosphorites can contribute to the uranium resources in many of these countries, if studied thoroughly. Other potentials and related evidence for uranium deposits in the MENA region are also summarized in this work with detailed maps, data, coordinates, and analyses.

This book will outline why the occurrences, deposition, and geological environments of uranium in the MENA varies from country to country. Another focus is given to phosphate-bearing deposits. For example, and scoping purposes, taking an average uranium content of 60–120 ppm, the Middle East phosphate resources would contain 1.2 million tons of uranium. The resource figures are, of course, much higher than for exploitable reserves that, nevertheless, run into the billions of tons. Another focus of the book will be on favorable geological environments for uranium in the Middle East, and their exploration techniques. A summary recognition criteria for the identification of uranium provinces especially in granitoids, volcanic and sandstone formation, and unconformity related deposits will be presented.

CHAPTER 1.4

Proposed metallogenic and tectonic classification of uranium ore deposits

The systematic review and field visits to uranium deposit in the MENA region motivated new classifications based on geological reasoning and metallogeny of uranium in the Middle East. The new proposed classification combines the geochemistry of uranium and its behavior in the earth's processes which is very versatile with the geological processes. As it is well known among geochemists, uranium has a crustal abundance of only 4 ppm, about the same as tin or arsenic; however it is found concentrated into uranium deposits in many different types of rock and geological environments, in concentrations of 20% or more. Uranium has two different oxidation states in nature. The plus four (+4) oxidation state is present in reduced conditions, meaning all magmas. In the more oxygenated surface and near-surface regions the plus six (+6) oxidation state is present in the form of the uranyl molecule, O-U-O "dumb-bell" shaped molecule, $(UO_2)^{+2}$. This cationic molecule hydrolyzes, very soluble in water, and transported in this manner. This is partly the reason for the diversity of uranium deposits.

The precipitation of uranium from surface or groundwater is necessary for the formation of several types of deposits. First, chemical reduction is the most efficient method of precipitation, reductants being carbon or sulfur in the forms of methane (natural gas), hydrogen sulfide, coal, organic materials, or shale. Independent of oxidation/reduction, the uranyl cationic molecule in water is attracted to a large variety of anionic ligands, including those of As, P, V, Mo, and where the concentrations are large enough, the solubility product is exceeded, and precipitation takes place. The fact that uranium mineralogy consists of over 100 different species that are members of many anionic mineral families attests to the diversity of uranium deposits.

As a result of these properties of uranium, there are many different uranium deposit types and subtypes, and many different deposit-type classifications. As countries have different geological characteristics, classifications tended to reflect national characteristics. The IAEA in 1988 provided a widely used classification based on uranium production by type. Significant problems exist in understanding genetic relationships between certain uranium deposit types, and the magmatic hosted deposits are most problematic.

Several deposit types are the name of a known deposit, such as Rossing, Namibia type, or the Bokan Mountain, Alaska type, with no relationship between them. It is the objective of the present chapter to provide an integration and synthesis of the various magmatic uranium deposits.

Stable blocks and mobile belts characterize cratons, and the mobile belts represent repetitive Wilson cycles. The boundaries between them, sometimes known as volcanic-rifted margins, are the regions where the magmatic part of the uranium deposition takes place. A model is presented integrating the following deposit types (1) metasomatic, (2) magmatic, (3) pegmatitic, (4) contact metamorphic, (5) hydrothermal veins, and (6) carbonatite, within a regional tectonic understanding. Another significant problem is demonstrated by a group of the surficial deposits. The origin of uraniferous coal, lignite, phosphate, and black shales is unified with the recent suggestion that the uranium comes from giant caldera ash eruptions from silicic large igneous provinces in adjacent extensional regimes. This implies that all four surficial regimes can be mineralized by the same event. This realization can be a significant global exploration tool.

CHAPTER 1.5

Geosystems model of uranium deposits

The formation and origin of uranium deposits is of interest for uranium exploration necessary for the sustainability of nuclear energy production. The proper organization of this knowledge is the most important. Prior models or classifications of uranium deposit types do not emphasize the geological relationships between deposit types illustrated by the uranium geochemical cycle. The proposed Geosystems Model/Classification of uranium deposit types starts at rift shoulders where thermal upwelling promotes anatexis, mafic intrusions, and A-type granitic batholiths, uraniferous caldera sand voluminous ash-flow tuffs. Numerous deposit types are in this igneous portion of the uranium geochemical cycle. Intra-cratonic sedimentary basins over geological time have provided multiple depositional opportunities. Several deposit types are outliers. The proposed Geosystems Model accommodates IAEA classes into the tectonic-U geochemical cycle–based interrelationships. New perspectives are the initiation of magmatism at rift shoulders, diversity of caldera environments, intra-cratonic basin template applicable over time, and surface uranium deposit types with accessibility to voluminous uraniferous air fall tuffs. The uranium to thorium ratio is a useful property for uniquely identifying different uranium sources, processes, and deposit types.

In 2015, Howari et al. (2016) presented a classification for the distributions of uranium deposits in the MENA regions. They reported that "these mineral systems can significantly contribute to our further understanding of the metallogeny of known and poorly explored deposits. This provides contribution to, and further enhancement of, current classifications and metallogenic models of uranium systems, allowing researchers to emphasize on unknown or poorly studied mineral systems found in MENA. The present study identified eight metallogenic types of uranium associated with: (1) the Archean rocks and intra-cratonic basins; (2) the Pan-African granites and rhyolites which are characterized by igneous activity; (3) Phanerozoic (Paleozoic) clastics, these deposits are the sedimentological response to Pan-African magmatism; (4) Mesozoic (basal) clastics type e.g., Nubia sandstones which are characterized by uranium minerals; (5) regional sedimentary phosphate deposits which are categorized as geosynclinal, or continental margin deposits, on the shelf of the Tethys Ocean; (6) Cenozoic Intra-cratonic Felsic Magmatism of the Tibesti and Hoggar, and the sandstone U deposits of adjoining Niger. These are similar to the Pan-African magmatism metallogenic, (7) calcretes, and (8) resistate minerals which are often enriched in rare-earth elements, sometimes including uranium. They are thus sometimes considered as U resources but poorly explored in the MENA region." This book describes in detail these deposits. Uranium provinces in Egypt; Islamic Republic of Iran; Iraq; Jordan; Saudi Arabia; Sudan; Syria; Turkey; United Arab Emirates, Oman, Qatar, and Kuwait; Yemen; Libya; Morocco; and Tunisia will be discussed and explained based on the common recognition criteria for uranium resources as well as the based on noted earlier geosystem models.

Reference

Howari, F.M., Goodell, P.C., Salman, A., 2016. Metallogenic evolution of uranium deposits in the Middle East and North Africa deposits. J. African Earth Sci. 2-1-2016.

CHAPTER 2

Geology of the Middle East and North Africa

CHAPTER 2.1

Introduction

The term the Middle East often means different countries to different people. However, geographically, and in terms of plate tectonics, the Middle East is the historical and political area of Afro-Eurasia as seen in Fig. 2.1.1. The Middle East, as defined in this chapter, includes Egypt, Oman, Yemen, Saudi Arabia, the United Arab Emirates, Bahrain, Qatar, Kuwait, Jordan, the fertile core of Iraq and Syria, southeastern parts of Turkey, and southwestern parts of Iran. Few countries in Africa share similar geographic and geologic history like that of the Middle East. Africa, which is the second-largest continent in terms of area, is divided into five main regions to facilitate the identification of each region. These regions include North Africa, South Africa, East Africa, West Africa, and Central Africa. It is difficult to define the precise dividing lines between the South, North, West, and East on the continent because the geography of the whole region bears more than one country within its boundary. The North Africa countries include Algeria, Egypt, Libya, Morocco, Tunisia, and Sudan. It is worth noting that the North Africa region is characterized by a strategic geographical location. It represents the link between the three continents of the world: Africa, Asia, and Europe. The North Africa region has the Atlas Mountains, which pass through parts of the countries of Morocco, northern Algeria, and some parts of Tunisia, and the North Africa region controls the Mediterranean climate due to its proximity to it.

This chapter will not present a systematic overview of the geology of the Middle East and North Africa (MENA) region but rather a snapshot on the important geological features and associated time period believed to be of interest to the mineralogical and metallogenic framework of the region (Howari et al., 2016). The region has significant diversity in terms of its geology, structure, and geochemistry. The age of the igneous, metamorphic, and sedimentary rock formations in this region ranges from Archean to Quaternary. The significant and important geotectonic features were

Uranium Geology of the Middle East and North Africa
DOI: https://doi.org/10.1016/B978-0-323-90992-1.00002-6

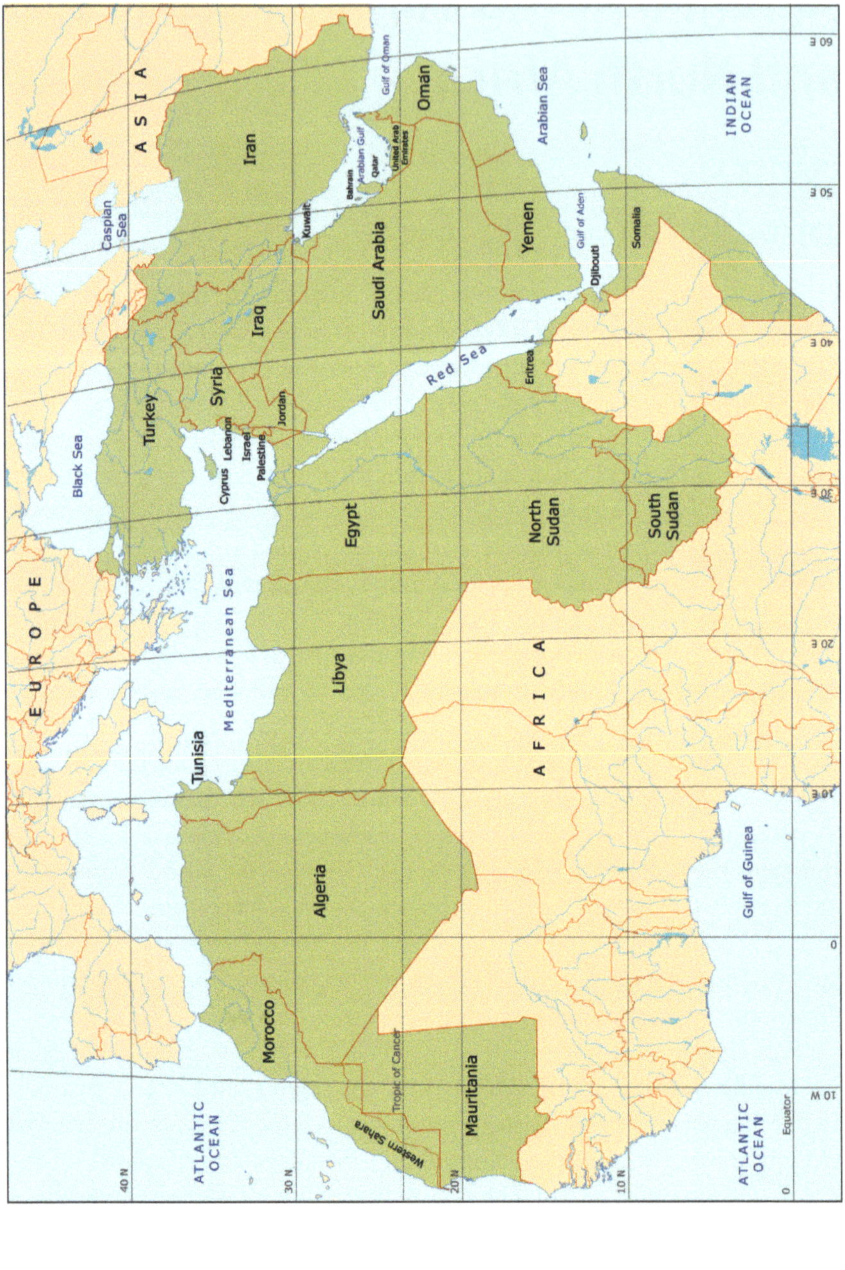

Fig. 2.1.1 *Map of the Middle East and North Africa.*

developed in three important time periods: (1) Proterozoic and subsequent Phanerozoic mountain-building events and activities; (2) the separation of the supercontinents Rodinia, Gondwana, and Pangea; and (3) during the conversion and diversion of the plates of the major ocean basins such as the Iapetus, Tethys Atlantic, and Pacific. "Three major tectonostratigraphic features are recognized (1) the Precambrian province that includes the African-Nubian Shield to the east and the West African Craton to the west, and the intervening Tuareg Shield and "Saharan Metacraton"; (2) the Variscan (Paleozoic) fold belt; and (3) the "Atlas–Alpine (Mesozoic–Cenozoic) system" (Bouabdellah and Slack, 2016). The focus will be on the crystalline basement, epicontinental sediment, geosynclinal or shelf deposits, Tertiary volcanic, epeirogenic movements, rifting, Tertiary orogeny, and Quaternary cover.

Since the geology of the world is not necessarily aligned with the political or geographic boundaries, references to other locations will be made occasionally since the MENA region in its current form was not developed in early geologic times as it is right now, in particular during late Protozoic time. The reference to other countries outside the MENA region will also happen since geologic extensions of some regional exposures extend or originate from places outside the MENA region. However, the geology of the MENA may be determined by the characteristics of the following regional geological features and events which will be discussed according to their geological time scale and relevancy to uranium occurrences.

According to Gass and Gibson (1969), geologically the MENA region is the area bounded to the Northeast by a Zone of Alpine deformation, and divided logically into three crustal units: (1) the Arabian Peninsula; (2) the Egypt–Sudan–Ethiopian part of the African continent, or sometimes called the Nubian segment; and (3) the Somaliland horn of the African region. Kent (1978) indicated that the geographical extension of the African Continent is the Central part of the Middle East. Further east and northeast, the numerous sedimentary and structural belts surround the bottom of this ancient center. Thus, it is useful to define the area into parallel zones. The first part includes the Arabian Shield and the shelf and sedimentary deposits descended under the Gulf; the second part is the coast of Iran and its associated foothills; the third zone includes the mountains of the Zagros; and the fourth is the high Plateau of Iran. A similar explanation of the geology of the major part of the MENA region into several geologic provenances, is shown in Fig. 2.1.2 and discussed in subsequent sections.

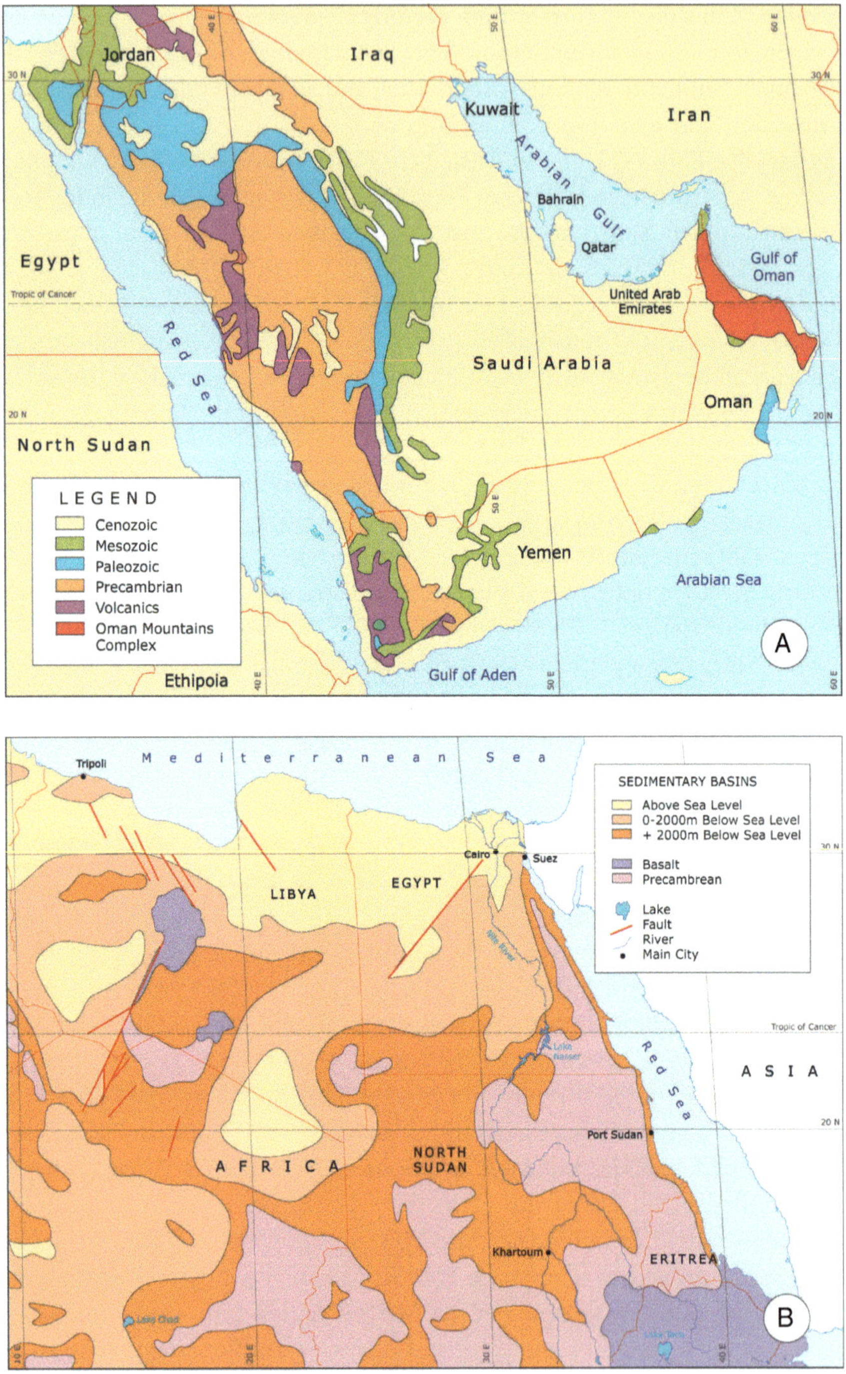

Fig. 2.1.2 *Geology of Arabia (A), and northeast Africa (B).*

CHAPTER 2.2

Morphology and structural features

According to Sadek (1977), geomorphologically, the Middle East can be divided into three provinces. The first includes the mountains and associated foothills network—comprising the range of Zagros and to the north, the Taurus range. The second includes a lowland in the middle area, mostly arid land covering major areas of Saudi Arabia, the western desert of Iraq, Syria, and Jordan. The third provenance includes the area on the west, with a highland shield, which comprises the southwestern portion of Saudi Arabia and Yemen. Depositional basin-wise, the geology of the Middle East can be classified into three main sedimentary basins: The Greater Arabian Basin, the Oman Basin, and the Zagros Basin. The regional tectonic framework of the Middle East reveals that the region is marked by several sub-basins, global peaks, flexures, and anticlines representing salt diapirism and deep-seated basement faults, which will be explained in detail in this chapter (Falcon, 1974; Kent, 1978; Jackson and Mckenzie, 1981, Beydoun, 1992, and others). The geomorphology of North Africa is dominated by well-formed, erosional features that developed as a polygenetic landforms due to successive events of regional uplift, and planation events that took place during the breakup of Cretaceous Gondwana (e.g., Burke and Gunnell, 2008; Partridge, 2010). King (1949) described relics of the breakup of Gondwana as a series of erosion features of various ages in southern Africa, including "the senile, continent-wide, African surface, later interpreted as a deeply weathered, low-altitude, largely flat, tropical plain formed at the end of the Cretaceous."

North Africa's main geographical feature is the Atlas mountain chain. This chain was formed by the same tectonic activities that developed the European Alps, which is the collision of the African and Eurasian tectonic plates. The basement rock of the major part of Africa was formed during the Precambrian supereon and is much older than the Atlas Mountains. The Atlas chain was also once attached to the Appalachian Mountains in North America. The mountain chains formed at the same time, during the formation of the vast supercontinent Pangaea between 540 and 250 million years ago. Later, when the Pacific Ocean opened, the two mountain ranges became separated. Meanwhile, the Middle East's main geographical features are the Alpine Mountain chain, which passes across Turkey, northeastern

Iraq, and through the Zagros Mountains of Iran to the Oman Mountains (Agar, 1987; Anneen, 1991; Neugebauer, 1995; and Nehlig et al., 1999). The mountainous area that passes through northern Iran's Lesser Caucasus and Alborz, enveloping the central Block of Iran, and the Arabian Shield and base, with its fairly undisturbed Phanerozoic sedimentary layer.

Figs. 2.2.1 and 2.2.2 show the structural divisions of the central portion of the Middle East. According to Nehlig et al. (1999), the composition of this region includes: (1) the central plain, known as the Nejd; (2) a series of desertland in the north, the Nefud with stony features; and in the south, the Rub' Al-Khali or Great Arabian Desert, with a deposit of sand measured at 600 feet underground; and between them, (3) the Dahna desert, a dry or marshy coastline with coral reefs on the Red Sea side (Tihamah); and (4) mountain ranges, mostly parallel to the Red Sea and Asir. The biggest, in Yemen, is Jabal Al-Nabi Sho'aib, 3666 m high.

Nehlig et al. (1999) described in detail the wide system of orogenic zones structured as a connection or arrangement of intersecting horizontal and vertical lines of predominantly left-sided strike-slip faults. The Nabitah Belt is aligned with a variety of ranges situated on the edge, along with molassic basins, and gneiss domes in the central part of the system. The meeting or intersecting points of the North–South and Northwest–Southeast networks in the central part of the shield outlines an s-shaped feature that represents their recent creation. Such features are not aligned with the arrangements of orogenic zones suggested by numerous authors (e.g., Stoeser and Camp, 1985), in which two tectonic events produced the Fault systems of Najd and Fault zone in Nabitah. Moore (1979) announced that a group of left lateral Northwest–Southeast strike-slip faults known as the Najd Faults are cutting the northern Arabian Shied. These faults are common in the creation of the discordant Jibalah carbonate platforms as well as the late Proterozoic Shammar Formation volcanic (Delfour, 1970).

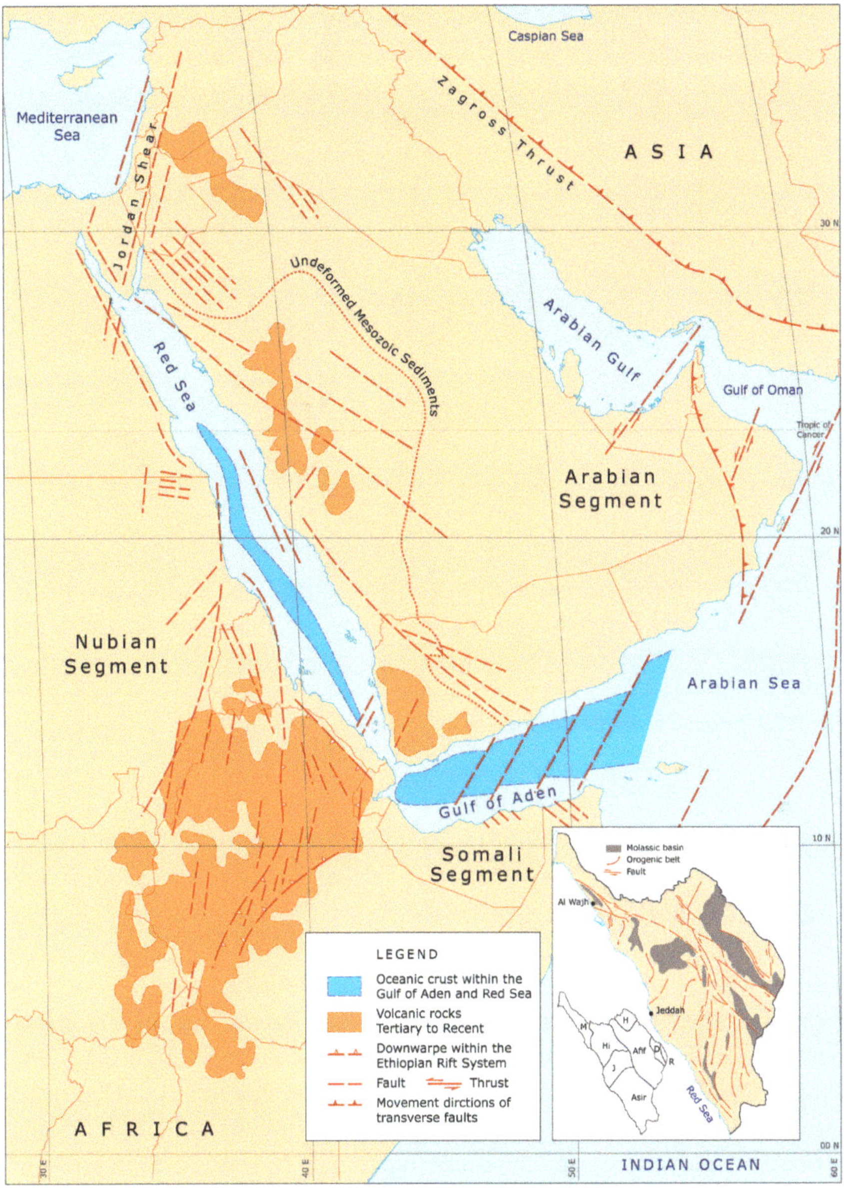

Fig. 2.2.1 ***Structural divisions of the central part of the Middle East.***

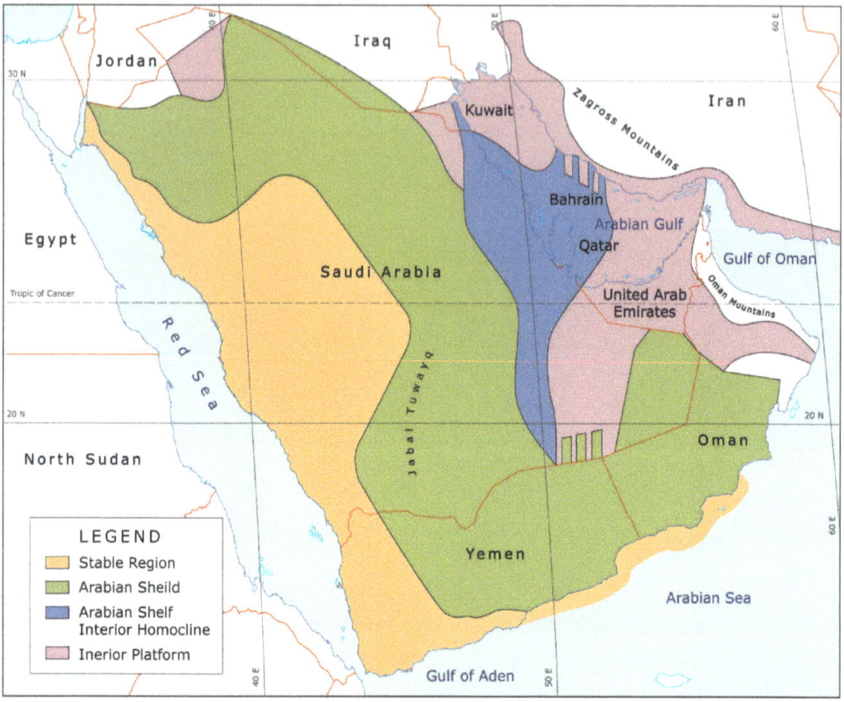

Fig. 2.2.2 ***Structural division of the Arabian Peninsula.***

CHAPTER 2.3

Geotectonic framework

The geological history of the MENA region most likely began with the creation of Archean–Paleoprotozoic continental crust, fragments of which outcrop in Yemen and West central KSA, and continue through the meso-Proterozoic, which assembled the supercontinent Rodinia, and continued into the Neo-Proterozoic, which was dominated by active tectonic rifting, formation, and distribution of ocean basins and the spread of animal and plant life, as well as the formation of the Pan-African basement, which was the result of dynamic Proterozoic orogenic evolution following collision and accretion onto a pre-Pan-African continent to the west of Egypt (Stoeser and Camp, 1985; Sage and Letouzey, 1990; Creling et al., 1994; Kesten, 2004). Kesten (2004) indicated that during this period, late orogenic magmatism became dominant, with the development of extensional basins.

The period that produced the beginning of the formation of the rocks in Pan-African was before 900 and 600 Ma for mega annum or "millions of years". It took place at the last stage of this Pan-African evolution. Several authors have re-calculated the time of the early tectonic compression process associated with these collisions. They also reported the presences of cal-alkaline, I-type granite batholith bodies and a subsequent extension process with A-type alkali granites. Some post granitoids mark the end of Pan-African activity (Creling et al., 1994; Kesten, 2004).

It is widely accepted that the major orogenies that influenced the MENA region include: (1) the late Proterozoic to early Phanerozoic Pan-African phase, which faded with the unification of Gondwana; (2) The main Paleozoic orogenies, the Hercynian and Caledonian. These bonded Pangea together but had a relatively insignificant impact in the cratonic areas (Abed, 1985; Walley, 1998; Burgoynea et al., 2005). Such Paleozoic mountain-building mechanisms were mainly the cause for up warping and the disappearance of the Silurian and Devonian in several parts of Arabia, Iraq, Turkey, and Iran.

The detailed geotectonic boundaries of the Arabian Peninsula are shown in Fig. 2.3.1. The Arabian Peninsula includes parts of Jordan and Iraq. Its boundaries to the west, southwest, and south are formed by tensional structures, the shear faults and grabens structure of the Gulf of Aden, the Red Sea, and the Gulf of Aqaba-Dead Sea rifts. The boundaries to the north and northeast are defined by the compressional features associated with the Alpine Taurus, Zagros, and Oman Mountains. It was divided into the exposed part of Precambrian portion or shield, and a cover of sediment throughout its geological history. The northern border is established by the Zagros collision zone, a mountainous uplift in which there is a continental collision between the Asia and Arabian plate. These areas also combine with the Syrian Desert without a strong demarcation line.

The eastern Mediterranean fringes in the vicinity, called the Levantine area, form a region where the African, Arabian, and the Eurasian plates meet (Neev, 1975; Walley, 1998). To the west, the Levant Platform makes the boundary of the NW part of the Arabian Block. The Red Sea Rift Valley is a broad, fault-bound depression breaking the top of the Afro-Arabian Shield. It is a feature that since the early Paleozoic period has been connected to the structural system of the Jordan-Dead Sea (Ben-Avraham, 1978; Abed, 1985; Gridler, 1987; Hirsch and Picard, 1988).

The region's Phanerozoic evolution primarily reflects its continental margin position and is distinguished by long wave transgression–regression

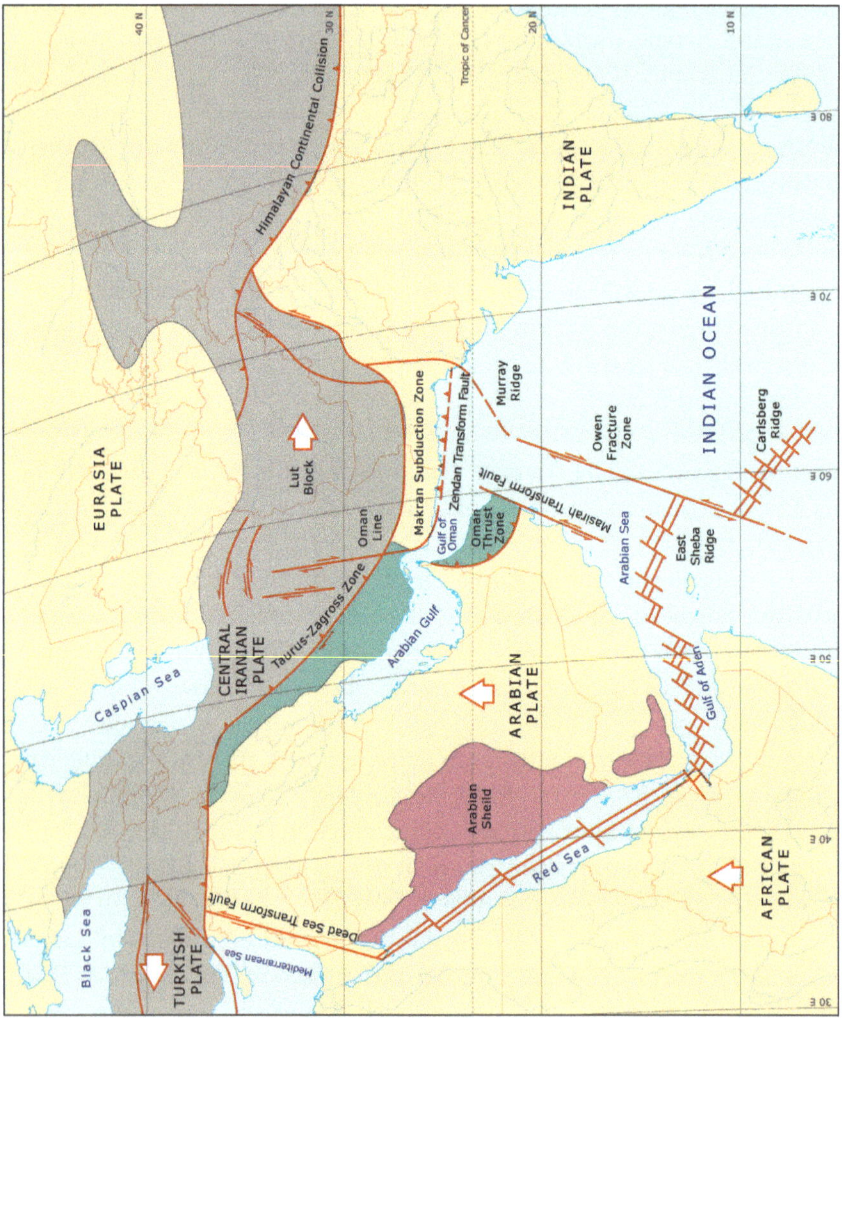

Fig. 2.3.1 *Compressional features associated with the Alpine Taurus, Zagros, and Oman Mountains.*

due to the Gondowana breakup. The area located south of the Proto-Teythes ocean during the Cambrian and Ordovician periods developed different geology due to the evolution of the Neo-Teths in the Levant region, which was affected by uplift and magnetism of various acid igneous rocks that are common in Levant and Western Sina. The Jurassic was a period of lithospheric extension and rifting, there was also a shallow marine sea over most of the Middle East. Lithospheric extension and magmatic activity concentrated in the early Cretaceous and was succeeded by the Syrian Arc deformation (Al-Sharhan and Kandall, 1986; Hirsch and Picard, 1988; Best et al., 1993; Kesten, 2004). This reformation resulted in shear zones, folds, and faults in the Sina-Dead sea area with significant Continental Cenozoic breakup and the Red Sea opening in the Oligocene, separating the Arab plate from the African plate (e.g., Cochran, 1983). The whole region was uplifted, and widespread tectonic activity took place (Ben-Avraham, 1978; Lovelock, 1984; Camp and Roobol, 1992; Kesten, 2004). One of the boundaries of the new plate resulted due to the formation of the Dead Sea between the Sina sub-plate and the Arabian plate. Is this the right word, or should it be a time period?

The Main Cenozoic orogenic activity that affected the Middle East is concentrated around Alpine-Himalayan belt, which includes mountain regions of Southern Europe and North Africa and extends eastward through the Middle East and India and into South Asia. Major deformation from the late Eocene to late Miocene occurred as the African and Arabian plates shifted northward against Eurasia. Pieces of the Arabian Shield, which dates to the Precambrian, are widely abundant in some Middle Eastern countries. Globally, Precambrian mineral resources yield most of the world's mineral resources. The Paleozoic geology of the Middle Eastern countries is a potential target for mineral resources, as evidenced from the various examples presented in this book.

Extension and features of Arabo-Nubian Shield

A significant geologic province for this book is the Arabo-Nubian Shield (ANS), a formation made from continental crust that is composed of mantle derived late Proterozoic rocks. The ANS is a member of the Orogen from East Africa (Kesten, 2004). It possibly formed through the period of the formation of Pan-African mountain-building territories or microplates at the northeastern boundary of the African Cratos, together with plutonism of calk-alkaline (I-type) rocks and metamorphism (Stoeser and Camp, 1985). Husseini (1988) suggested a range of 620–580 Ma for a related collisional tectonic system that was replaced by extensional tectonics, accompanied by rapid erosion that took place and removed 8–12 km of crust in 20 Ma (Garfunkel, 2000). Igneous activity was dominated by alkaline, and A-type rocks during this time (McKenzie et al., 1970; Kesten, 2004).

The ANS is an example of crystalline of rocks of Precambrian age on the Red Sea outermost's strip. The age of the hard rocks is mainly Neoproterozoic. The ANS includes Palestine/Israel, Jordan, Saudi Arabia, Egypt, Eritrea, Sudan, Ethiopia, Somalia, and Yemen. In the Arabian Desert and Sahara Desert, the ANS is visible in the north, and in the south, it appears in the Ethiopian Highlands, Yemen Highlands, and Asir province of Arabia (De Sitter, 1962; Nehlig et al., 1999; Johnson and Woldehaimanot, 2003). According to Johnson and Woldehaimanot (2003), the ANS in the north consists of a huge collisional region called the East African Orogen. This region of a Neoproterozoic time's collision developed when West and East Gondwana came together to form the Gondwana supercontinent. The most violent aspect of the collision happened in what is now southern Africa, where Tanzania consists of older crust. Mozambique and Madagascar had been removed to form the Mozambique Belt. The Pan-African orogeny at the end of Neoproterozoic periods was responsible for these great geological changes in Africa. Older Archean and Paleoproterozoic crustal materials are found in Egypt to the west of the Nile, Southeastern Arabia, eastern Ethiopia, and Yemen. Further to the north and east, the ANS is influenced by the development of the strike-slip faults in the Najd *Fault* System. As the impact progressed and the crust began to thicken, the structure of igneous rocks became markedly more advanced. Extreme erosion, likely from a continental ice cap, emerged at this period of both magmatic and tectonic activity, which finished when the sandstones of Cambrian ages were formed about 530 million years ago. Many of the Neoproterozoic-age crystalline rocks cover the areas surrounding the Red Sea (Fig. 2.3.2).

Fig. 2.3.2 ***Neoproterozoic-age crystalline rocks cover the areas surrounding the Red Sea.***

CHAPTER 2.4

Regional geology and associated chronology perspective

Precambrian basement

Near the end of the Precambrian, large parts of the MENA region were influenced by the Pan-African movements, as noted earlier. This caused a tectonomagmatic episode in which Archean to earlier Proterozoic rocks were removed, producing deformation and re-metamorphism to facies of granulite attributes. The crystalline basement in the Middle Eastern was exposed in formations as rims to North Africa basins of epicontinental origin and is mainly composed of sedimentary rocks. The size of these outcrops declines in the direction of the North. They form the large blocks of Tibesti, Uweinat, and Ahaggar descending from the Sahara, and the long range of hills and mountains comprising most of the Eastern Desert of Egypt (Figs. 2.4.1 and 2.4.2). These basement rocks are found in Egypt, Sudan, Yemen, Jordan, and Saudi Arabia. For example, Precambrian rocks in Jordan are exposed in Aqaba, Wadi Araba, Ras El-Naqap, and Wadi Rum. Cambro-ordovician (Wajid) sandstone outcrops also exist in Yemen (Gass and Gibson, 1969; Haberland et al., 2007).

The basement in Egypt is found in the Eastern Desert and consists of a rugged basement massif, running parallel to, and at a relatively short distance from the coast. To the south, these basement rocks (or mountains) are much broader than northward, where they disappear at the latitude of Wadi Araba. These mountains have a dentatae profile and are constituted of the Arabo-Nubian massif's basement rocks. A linear structure running north–south, they are abruptly cut off to the west at Wadi Qena. The chain is drained on both sides into the Red Sea and Valley of the Nile. In the same way as mentioned above, it is covered by a mantle of post-Precambrian sedimentary rocks dipping generally to the east and to the west. Southward, Nubia Sandstone rests unconformably over the basement rocks in the vicinity of the Nile Valley. This Formation is followed by upper Cretaceous (Cenomanian and Turonian deposits, variegated shales, and chalk).

Beydoun (1964) demonstrated the existence of a young Precambrian crust in South Yemen with a history essentially the same as in KSA. This

Fig. 2.4.1 ***The large blocks of Tibesti, Uweinat, and Ahaggar descending from the Sahara.***

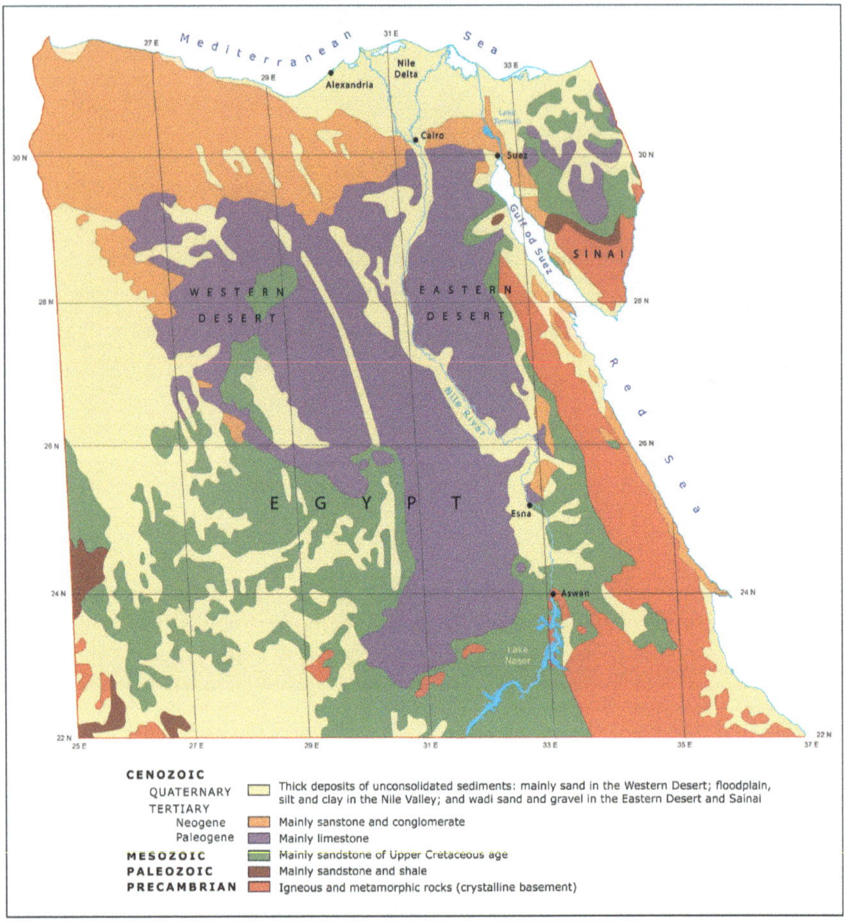

Fig. 2.4.2 ***The location of Tibesti, Uweinat, and Ahaggar descending from the Sahara, and the long range of hills and mountains comprising most of the Eastern Desert of Egypt.***

crust spreads much further east in the far south of the Arabian Peninsula than in KSA. For example, in Egypt, anything eastern direction from the Nile belongs to the Arabo-Nubian Massif. Westward, there is no evidences of Precambrian rocks outcropped north of latitude 23°25′. Southward and along the Sudanese boundary, a discontinuous belt of still almost unknown Precambrian rocks stretches westward toward the Precambrian formation at Jebel Uweinat, at the junction of Egypt, Sudan, and Libya. The basement beds experience cuts due to the Red Sea rifting especially in the west of KSA and Yemen and they are less visible in around Aqaba and Sinai

(Dubertret, 1970; Quennell, 1956; Quennell, 1984). In this part of Asia, which is buried under thick sediments, they do not outcrop further East or North. Kroner et al. (1987) categorized the basement rocks into four major rock groups: (1) older shelf-sets, (2) assemblages of arcs, (3) suites of ophiolite, and (4) intrusives of granitoids. The Afro-Arabian Shield's ophiolite sequences exist as allochthonous and usually dismembered ultrabasic and fundamental structures, including serpentinite thrust mélanges and ophiolitic mélanges, and fault-bounded inliers.

Infracambrian

There is a dense sedimentary layer beneath the fossiliferous Cambrian in many parts of the world, and the Precambrian basement rocks are often unconformable overlying (Sadek, 1977). Pruvest (1951) called these middle beds the "Infracambrian." Several reports, for example, demonstrated a stratigraphic section nearly 2000 m thick in southwestern Turkey, consisting of elastic deposits named as the Derik Formation. This portion is 800 m of red beds and polygenetic conglomerates of coarse sandstones. These outcrops are called the Sadan Formation and cover the top of the formation. Flugel (1964) confirmed the occurrence of volcanic acids in southeast Turkey's Infracambrian. They reflect magmatic eruptions during the "Assyntic Orogeny" that occurred before the Precambrian ended. The period between the Precambrian basement and the fossiliferous Lalun Sandstone, Cambrian age formation in Iran covers a dense sequence of micaceous variegated shales, purple sandstones, stromatolitic limestones, and cherty dolomites. The Infracambrian deposit, however, is believed to stretch to the Arab Gulf, and other area of the Middle East (Sadek, 1977).

Cambrian

In most parts of the MENA region, from Cambrian to the Cretaceous, the formed crystalline rocks were almost covered by Phanerozoic sedimentary rocks until uplift and erosion associated with the separation of Africa resulted in exposures of the Arabian Shield. Lower Cambrian to Middle Cambrian, major sandstone deposits were formed due to regression and transgression of the Tethys (e.g., Cambro-Ordovician (Wajid) sandstone in Yemen, Wadi Araba sandstone, and Nubean Sandstone). The Cambrians, on the Mediterranean coast of the Turkish south, are formed by a richly fossiliferous, yellowish shale with trilobites. They have an eastern Cambrian age for basal conglomerate and quartzite–sandstone beds, also containing trilobite fragments in the Mountains of Amanos. Kamen-kaye (1971) reported that the Cambrian sediments are not evident in "Northern and Central Parts of Turkey." The well-known Mardin exposures to the southeast expose a probable 1900 m of sandstone, silt, and shales. In another part of the MENA region, Sadek (1977) recorded the Cambrian. He recorded that in Iran, Cambrian rocks are composed mostly of two main units of lithology. On top are often richly fossiliferous shales and calcareous rocks, reflecting the upper unit, while the basal unit consists of sandstones (Lalun Sandstone Formation). The thickness of the upper unit ranges from 20 to 2000 m. The fossils found—mainly trilobites—usually indicate the Middle and/or late Cambrian period. The basal unit (Lalun Sandstone Formation) is made of purple to red crossbedded sandstones, arkose to quartzite sandstones, and ranges in thickness from 500 to 1000 m.

The deposits of Cambrian age in the north of KSA are represented by the basal portion of a sandstone sequence with 600 m of thickness, which is known as "Saq Sandstone Formation" (Sdek, 1977; Kent, 1978). The basal portion of the Saq and similar sandstones (Jordan's, Ram, Umm Salim, Quweira, and Siq) overlap the "Precambrian crystalline complex" strongly. The top is delineated at the sandstone contact with the "overlying graptolitic shale" (Ordovician age Tabuk Formation). Most of the Saq Sandstone is trustfully attributed to the Cambrian (Sadek, 1977). In Egypt, Said (1962) recorded sporadic occurrence of early Paleozoic rocks in the Egypt's NW desert. The upper Cambrian–lower Ordovician is assigned to the well located in the Gibb Alia and exactly in NNW desert of Egypt.

Ordovician

The Ordovician exposures in the Middle East are described by Turkunals (1953), Van Bellen et al. (1959), Binkmann (1971), and Sadek (1977). These exposures in northern and southern parts of Turkey were identified by Van Bellen et al. (1959). The Ordovician deposits in southern Antolia are composed primarily of marine siltstones and sandstones. In the northern part of Anatolia, they are exposed as a red, coarse, elastic chain, consisting of conglomerate quartzite sandstones and greywackes thicker than 8000 m, resting over older crystalline rocks. The Ordovician deposits in southeastern Turkey consist mainly of varicolored intercalations (800 m thick) of shales, sandstones, and siltstones. In Ricgar Dag, north of the Turkish–Iraqi border, at the southeastern edge corner of Turkey, Turkunals (1953) reported the Cambrian–Ordovan series

According to Sadek (1977), along the far northern edge of Iraq, the Ordovician rocks are visible. They are divided into three stratigraphic components: Chalki Volcanics, Pirispi Beds, and the Quartzite-Shale formation Khabour, in order from top to bottom. Geomorphologically, the sediments described above are exposed in Geli Sinat (the extreme western side of the Iraqi thrust zone), Khabour River, Dershish Village, and Ora Village. Ordovician rocks form the center of a small anticline traversed by the Geli Sinat River along a NE–SW trend at the Geli Sinat region. They are faulted northward against Cretaceous rocks and along a "Major Thrust Fault" segment that runs in a WNW–ESE trend. At Dershish Village, approximately 8 km east of the Geli Sinat region, Ordovician age strata form the center of an elongated anticline that trends almost parallel to the main thrust fault (WNW–ESE). In this area they are faulted eastward against late Paleozoic rocks, and westward by a Northeast–Southwest fault against Mesozoic (Triassic) rocks, respectively.

Chalki Volcanics were reported at the top of Pirispiki Red Beds (Van Bellen et al., 1959) to the northeast of Dershish Village. For about 80 km east from Dershish Village, Ordovician rocks are present in the area between Khabour River and Kaista Village. They cover an area of approximately 85 km^2 which forms the center of the Kaista anticline, which trends ENE–WSW. They are surrounded in all directions by late Paleozoic and Mesozoic strata, except for the South part of the system where the over thrusting of late Paleozoic strata vanishes. Sadek (1977) noted that in many locations in the central part (Sanand-Sirj) Ordovician and early Silurian deposits are totally absent, although they are well formed in the east, north, and northwest (Takab and Azerbaijan). The early Ordovician sediments are characterized by sandy–silty shales ranging in thickness from 400 m to 1800 m in the Tabas region northeast of Iran, and in the Alborz Mountains in the north. On the northwest of the Takab region, these deposits transform laterally into well-bedded, around 200 m thick, calcareous stone. The Ordovician and Silurian rocks are apparently absent in the central part of Iran. The Ordovician formation in northern KSA is characterized by the Tabuk Formation's lower section (Ordovician–Devonian), which consists primarily of sandstones, which is intercalated in thin layers of shale, around 817.6 m thick.

Epicontinental sediments

Beginning with a broad thickness of lower Paleozoic Sandstones, the epicontinental sequence can reach 2000 m in thickness. The sandstones of Cambrian age seem to have spread over very larger regions as a shield. There are several marine deposits in the succeeding Paleozoic and Mesozoic, but sandstones, which appear to become arkositic, continue to predominate. According to Powers (1966), in Saudi Arabia there are about 2300 m of Cambro-Devonian sandstones where the sedimentary sequence falls very slightly off the Precambrian shield. Intercalated marine (especially Jurassic and Cretaceous) sediments succeed them, but with some sequences of sandstones.

Kent (1978) also reported that the sediments immediately above the Crystalline Shield are the Paleozoic elastics, followed by Mesozoic and Permian rocks, gradually dipping North–East. At outcrop, old Paleozoics are unsettled, but it appears that they are dislocated to further east—maybe by faulting before being beveled by the Permian seas. The sediments outcrop in parallel escarpments, some of which are continuous for 500–1000 km. This outcrop trends South–East from southern Iraq across Saudi Arabia, hooking north–east in parallel to the Arab Sea coast in Southern Oman and Dhofar. Northeast of the Homocline, broad contiguous basins extend to the Gulf, mostly of Tertiary rocks covered by sand and gravel. Gentle folding is only established in the third region. Belonging to these sediments is the Nubian Sandstone. The subsequent paragraph emphasizes this deposit due to its relevancy to uranium occurrences. The paragraphs are adapted mainly from Mckee (1963). Usage of the word Nubian has spread widely, geographically as well as stratigraphically. The term has been used for rock from Cambrian to Cretaceous in age in sandstone bodies in most North African countries and on the Arabian Peninsula. Shukm and Said (1944) address several ways in which the word Nubian was used to characterize the rock units in Egypt. Bentor (1960) sum up the complex of uses in Israel and Jordan. Two main trends have developed, the first was to reduce the Nubian name to Cretaceous sandstone of the age; the other was to use the word for facies and use it in all areas, irrespective of age, where there is sandstone with some general characteristics.

Mckee (1963) claimed that where the Nubian is wholly of Cretaceous age and rests on a peneplain pre-Cambrian floor, it is composed of a collection of mostly unfossiliferous sandstones, yellow, brown, or buff, which are clearly cross linked and are very fine gravel lenses or spread over the forest surfaces. Quartz grains are usually mild to substantial and mostly fine to medium grains with individual sandstone units. Many of the characteristic features of the Nubian type exist in other sandstones called Nubian but also substantial differences can be seen in critical studies. Maybe the two features of the classification of this sandstone are much more common than any other: (1) the broad and relatively uniform distribution of clean sand units, and (2) the abundance of prominent cross layers within this sandstone.

There are several hypotheses concerning Nubian-type sandstone genesis. Its composition was assigned to difference marine environments in the fluviatile, eolian, lacustrine, estuarine, then near-shore and beach regions. Certain forms of evidence applied to the Nubian, such as the shape of ripple marks and the occurrence of concentrations of gravel, that discern wind from water as precipitating agents, but may not be used to differentiate between marine and fluviatile deposits. Sedimentation in more than one continental, marginal, or maybe, locally, marine climate may also be caused by characteristics such as fossil plants that are "embedded in the sandstone." There are also several problems in determining the origin of this sandstone conclusively and these are exacerbated by the possibility that more than one deposition agent could have been involved in a locality at any given time.

The Fezzan, Libya

On the north boundary of Murzuk Basin in the Fezzan province of Libya, early Cretaceous age rock, generally referred to as Nubian Sandstone, and the rock below, are classified as Continental Post-Tassilien Group and assumed to be of Carboniferous to early Cretaceous in age (Burollet, 1960). They form hundreds of kilometers of north-facing steepness or rim stretching. The rock is predominantly sandstone, 1–8 feet thick in bedding sets but it includes shallow mudstone at a few horizons. Mudstone is most common within the region studied at three Mountains in the northeast. Petrified wood is present in Libya's Nubian and related sandstones, but it is not plentiful. The presence indicates a continental deposition. The most prominent characteristics of the Nubian Sandstone in Libya, and perhaps the most important from the point of view of understanding genesis, are primary structures. They include different forms of stratification and cross stratification, which have been engraved prominently on weathered surfaces. Some flat cross-stratification systems near Bint Biyyah have separate bounding planes, which dip 19°–14°. Between them, cross strata dips about twice these amounts in the same direction. Ripple marks were noted at several locations in Nubian Sandstone flat beds, and particularly in areas at Wadi Ajial, in the southwest of Sebha Oasis. On one horizon, 8 inch was averaged from crest to crest, and ¼ inch deep: on the two other horizons, 9 inch in decent grade, 90% in grades 8 or 4, badly sorted, and 90%in grades 5 or 6 inches.

Kharga-Dakhla Oases, Egypt

Nubian Sandstone is dated in the Kharga-Dakhla Oases region of Egypt's western desert from early Maestrichtian (late Cretaceous) based on fossilized plants. Such reservoirs in the region of Ain Amur, 40 km northwest of E1 Kharga, are 9.70 m thick, with mostly marine mudstone in top 70 m and sandstone on the bottom 200 m, with shallow, mudstone partings. There is little conglomerate in the Nubian Sandstone in this region. In cross beds of the sandstone no gravels were noted, either dispersed or concentrated, and in one section a single horizontal conglomerate of jasper and chert was noted as just a few inches thick. Lack of conglomerate in this area's Nubian Sandstone contrasts with cross beds and very gravelly beds that shape almost all the area of Khartoum to the south. Significant evidence suggests that in this region the Nubian Sandstone has a fluviatile and floodplain origin. The source of water is shown by (1) tiny cross beds with planar ups and downs, (2) mudstone bed with associated beds, (3) "conglomerate beds," (4) wide "cusp-type ripple marks" (McKee, 1963).

The Nubian Sandstone of the Aswan district

Attia (1955) defines the Aswan district's Nubian Sandstone in the eastern desert of Egypt as varying from 70 m to 162 m thick and divided into three parts—a lower sandstone unit, claystone, and conglomerate; a middle sandstone with oolitic hematite and claystone; and the upper unit mostly comprised of sandstone but with some claystone. The Nubian Sandstone to the east of Aswan includes small granules and some quartz pebbles, the cross-stratified sandstone is clean and largely non-conglomeratic. Horizontal conglomerate beds occur between sets of cross-stratified sandstone at certain horizons west of the Um Barmil region, with gravel ranging up to 1 inch in diameter. The Nubian Sandstone's origin in the district of Aswan is evident, but different geologists have interpreted its source in various ways: marine, continental, estuarine. The central unit, whose invertebrate fauna, freshwater and marine mollusks, iron deposits, and its oolite structure, have no characteristics that indicate a marine climate.

Khartoum region, Sudan

The age of the Sudan Nubian Series is late Cretaceous (Andrew, 1948). It is 150 m thick, and the lower unit is sandstone and conglomerate, the middle unit is mudstone, and it has an upper unit of sandstone, which is subdivided into three classes by Sandford (1985). Only basal beds are exposed near Jebel Mudaha and Jebel Merkhiyat, north–west and southwest of Khartottm, respectively. The middle unit of Jebel Aulia Dam, about 300 m above the pre-Cambrian one, is represented and at Jebel E1 Ghar, and Jebel Amuriya, east of Khartoum, the upper part of the section forms the outcrops.

The sandstones in the Nubia series show poorer grade than any other studied from in the region. Most probably the poor classification is due to the proximity to the source. The rock is usually vary conglomerated in basal units (Jebel Mudaha) with a diameter of 6–7 inches in the gravel, and 1/2 inch to more in the 80 m of base (Jebel E1 Ghar). This rock is strongly conglomerated. The sedimentation system in the Sudan region is clearly non-marine. Gravel excess, tabular-planar cross bedding, and difficult classification exclude eolian processes. Fossil plants, lack of winnowing data, and relatively poor sorting render it unlikely to have originated on or near a strandline. Finally, the dominance of tabular-planar system is an indicator of deposition of the forest beds in a standing water body. In general, it is not understood why the forest dips are lower than elsewhere in the Nubian Sandstone. Alluvial plains of delta cones tend to be possible deposition habitats.

Wadi Araba, west side Gulf of Suez, Egypt

In this region, Cretaceous age Nubian Sandstone was briefly examined to compare it to the Cretaceous rocks at Kharga Oasis, Aswan, and elsewhere to the south, in a series of ridges along the Wadi Araba's southern side. The sandstone to mudstone proportion is much lower in this field than is studied elsewhere, where sandstone units 5–15 feet thick are divided between 30 and 50 feet in thickness by shallow mudstone units. This decline in the percentage of sandstone, as calculated elsewhere, is fully expected to go away from the direction of detrital source, if the rocks have approximately equal periods.

Sinai Peninsula, Egypt

Carboniferous structures are applied to a sandstone dominantly cross stratified, which is similar to Nubian in many respects on Egypt's Sinai Peninsula. A series of rocks was broken down into three parts (Shata, 1955): (1) a lower range of unfossiliferous sandstone and a shale, with a thickness of 188 m on Precambrian rock; (2) a small variety of marine fossiliferous limestone, with a thickness of 40 m; and (3) a 150-m-high sandstone band, with an abundance of plant's fossils and carbonaceous matter.

The Sinai Nubian-type sandstone was examined in three locations near Um Bogma with eight horizons, each having a range of different characteristics. The Wadi Buda, where most of the lower part of the stratigraphic portion is visible and recognized as six sandstone units reflecting various environments. This includes marine as well as continental environments. High section, distinctive sandstone units were studied at Wadi Naga E1 Masaud and in Wadi Aou Arta's mouths.

The highest level of exposure is a smooth, fine-grained sandstone that has a wide cross-layered region at Wadi Naga E1 Mesaud and Wadi Buda, 78 m above a horizon of dolomite. Wadi Aou Arta has the Sinai Carboniferous rock's thickest sandstone band. It consists of a cross-stratification trough and displays a pattern of moderately broad festoons in the cut sections with a north–south orientation. Many of them are 15 or 20 feet wide, 7 or 8 feet deep, and plunging east. Slumping on one festoon's side resulted in a near-vertical-attitude local structure. The Sinai Peninsula's Carboniferous sandstone clearly represents not one but a variety of different condition of deposition in its various parts, ranging from beach to fluviatile and likely eolian. The different units reflect different regressive stages throughout this region (McKee, 1963).

Vicinity of Jebel Ram and El Quweira, southern Jordan

In 1914, Blanckenhorn described cross-stratified sandstone as Nubian Sandstone in Jordan, ranging in age from Cambrian until Middle Cretaceous, and later was called Petra Sand. The Ram Sandstone exhibits several features which have the same characteristics as the Quweira basal portion. Similar to that, several beds contain scattered pebbles, some with 8 inch diameters in Jebel Ram but usually with as much as 4–8 inches at the locality of Ras en Naqb. Nonetheless, a good proportion of lithological sets in each region are made up of clean, uniform, quartz sand. In the case of Rams, approximately one third were sets of 10–80 feet in length and several sets of 5-foot sizes were found near Rasen. In other words, a significant amount of cross stratification is of a table-planar type, which has sets of between 1 and 8 feet thick, but unlike Quweira Ram also contains multiple wedge planar sets of a bigger scale (McKee, 1963).

The recumbent folds of Ram Sandstone are not rare in table-planar sets anywhere visited and some very twisted cross-country beds have been found in Ras en Naqb with many folds. At the latter position and at Petra forward-dipping sets are not uncommon to be with 5°–7° of tabular-planar cross strata that is 18° dip to 24° dip. Within cross beds, the structures look like cross beds. All of these are considered as a diagnostic aqueous deposition of the mentioned textures and structures in Quweira and in most Ram Sandstone. In Ram, the broad and truncated-coiled structure as well as the fine, evenly grained composition of certain sets were often described as either wind or water deposition characteristics.

West of Salt, northwest Jordan

Attributed to the lower Cretaceous, the Kurnub Sandstone was investigated along the Arda Road in north-western Jordan between the Jordan River graben and the Salt town. The cross-stratified sandstone is much smaller than the most other Nubian-style sandstones. The sandstone is subtly grained and well sorted, without rocking, as seen in a typical sample from Jarash near Salz (McKee, 1963). The most common stratification in the Kurnub is tabular-planar, but there are also some small troughs. Nearly every set of dipping cross beds are 1–8 feet thick. Around two thirds of the cross beds dip between 20° and 80°, while the rest dip 10°–20° compared with other Nubian systems. This is probably the most distinguishing feature, other than its specific unit association (McKee, 1963).

Ramon Valley in Negev, Israel

In the Middle Jurassic era, Nubian-type sandstone was found in Ramon Valley in Israel's Negev desert. The rock here is mostly coarse- to very coarse-grained and has a moderate to good form. This sandstone seems anomalous since it is relatively well known that the principal Jurassic sea lay to the North–West and transgressed to the South–East, but the presence of land plants and the form of crossbedding suggest the deposition of these sites is continental, and a western movement is deduced. Probably the tabular-planar structures are those of a bar traversing a lagoon (Fischer, 1961), in which case the dips will be shoreward (Mckee and Strerrett, 1961). This topic demands further research. The Makhtesh Hathira's lower cretaceous rocks in Israel contain Nubian-type sandstone units between the units containing 482-m-thick layers of marine fossils (McKee, 1963).

Geosynclinals/shelf deposits

The Tethys Ocean and Sea bordered the Middle East area on the North, into which products of terrestrial erosion passed and interdigitated with sediments of clear marine origin (Kent, 1969; Pufahla et al., 2003). As the Tethys in the Tertiary narrowed and shallowed, deposits of evaporation became important and deposited Fars formation. Such thick sequences of sediments form an unusual northern limit to the region being considered here. Others are folded softly, as in Libya's Djebel Akhdar; others are bent and less folded in Palmyrean chains than in Syria and Lebanon shear folds; some are profoundly folded and thrust, as in the Zagros and Oman mountains.

According to Pufahla et al. (2003), the margin of the Tethyan passed through an intermittent period of tectonism accompanied with the northward migration of African and Arab plates into Eurasia during the late Cretaceous. Such bursts culminated in a widespread surface sediment syn-depositional fold, creating an undulating topography of the sea floor and consisting of sub-basin and intrashelf swells (Kolodny, 1967; Reiss, 1988; and Abed and Sadaqah, 1998). The most famous episodes of these in the east Mediterranean formed the Syrian Arc, an S-shaped fold belt (Freund, 1965). The Syrian Arc consists of a sequence of asymmetric plunging anticlines tending northeast to southwest with amplitudes of up to 700 m and wavelengths of about 20 km. The deformation of upper Cretaceous phosphorite on this tectonic front coincided with the deformation of the Coniacian and into the Miocene (Abed, 1988). As a result, phosphate deposits are widely dispersed in the MENA region especially in Tunisia, Jordan, Morocco, Egypt, Algeria, Iraq, Palestine, Syria, and Israel.

The rocks were deposited on a stable marine-to-fluviatile shelf in the Arab platform. The rise and collapse of arches and basins, the displacement of fault blocks, and the migration of shores across the coast have led to cross sections and facies of sandstones, siltstones, carbonates, and evaporites (salt basins) that define this region is Phanerozoic. In the initial intracontinental rift which formed, with ongoing spread into the current sea basin, "Cenozoic sedimentary," evaporitic, and small volcanic rocks filled the Red Sea basin (McBride and Barzan, 1990).

Tertiary volcanic

The volcanic activities in the Middle East have been described by several researchers such as Bender (1968); Camp et al. (1987), Dubray et al. (1991), Mouty et al. (1992), and Shaw et al. (2003). Rifting is indicated by basalts (some with small acid extrusive/rings), and their thickness in Ethiopia and neighboring areas are large. They are found in several places along the rift border, as in Syria/Jordan's Hauran and Djebel Druse. They also develop in the west, such as the Sudanese Djebei Marra and the Libyan Djebels Soda and Haroudj. Such basalts on the plateau may be major aquifers. Volcanic activity has persisted through history. Flows can also contain land and sea deposits that can also be aquifers.

Shaw et al. (2003) reported Oligocene flood volcanics and the probable presence of a mantle plume in the Afro-Arab tri-point junction of Ethiopia, Yemen, Eritrea, and Djibouti (Chazot and Bertrand, 1993; Baker et al., 1996; Stewart and Rogers, 1996; Pik et al., 1999). But magma and the geodynamical impacts of the various volcanic intraplate fields are less evident on the Arab plate (Camp and Roobol, 1992). In the last 15 Myr, the west margin of the Arabian Peninsula (i.e., Syria, Jordan, Saudi Arabia, Yemen, and Israel) has had a widespread "intraplate volcanism" occurrence. This intraplate volcanic system dates from the break-up of Arabia and Africa, and from the opening of Yemen and Ethiopia's Red Sea (Camp and Roobol, 1992). While very extensive, the volcanism of late Cenozoic intraplates is volumetrically insignificant, as is Oligocene Ethiopia and Yemen (12,000 km^3 vs. 350,000 km^3) occurrences, which may represent lithospheric temperatures and/or lower mantle than those responsible for massive volcanoes.

For instance, the Afar plume was connected to Oligocene volcanic flooding in Ethiopia and Yemen and contributed to the recent Pliocene–Recent intraplate volcanism in Yemen (Show, 2003). It was suggested that small amounts of lithospheric extension had released and caused the melting of flaky mantle, the Afar plume was previously hydrated and metasomatized (Baker et al., 1998). Yet magma genesis and the geodynamic-effected volcanism are not so clear in northwestern Arabia as far as the Afar's plume is concerned. For a partial melting of dry mantle, the typical mantle potential temperature of 1280°C is appropriate as is the extending factor (initial/final lithosphere thickness) of 2–5 (McKenzie and Bickle, 1988). In comparison, ancient, shallow mantle plumes of Saudi Arabian and Levantine alkali were suggested by Stein and Hofmann (1992) and Altherr et al. (1990). Stein and Hofmann (1992) hypothesized that in the sub-continental lithosphere an old (>200 Ma) "fossilized" plume head was deposited. Wide areas in Yemen are filled by extrusive Tertiary and Quaternary volcanic rocks. They are known as Yemen's volcanism. They include rhyolite to basalt compositions.

Epeirogenic movements

From the basement history, rises and falls in relation to the seabed have been evidenced by sea incursions far and wide in North Africa and Arabia's vast land masses. The basement has also deformed itself into the African basin-and-swell system, with sediment bonding down to a thickness 6000 m. Some of the swellings can be attributed to plate tectonics structures and were mostly found in Tertiary volcanic eruptions. According to Immenhauser (2005), the largest carbonate layer occupied large portions of the Arab craton from the middle of the Permian to the earliest Turonian (Sharland et al., 2001). During the Albian period, this platform was formed by several distinctive paleo-depositional belts. The Arab–Nubian Shield in the southwest, ranging northeast to a shallow water carbonate shelf, ranges from alluvial to coastal deposits.

Rifting

The region is divided into two by Suez Red Sea-Gulf and associated rifts, which stretch from Aqaba to the Ghab of Syria by way of the Dead Sea. Several other subsidiary rifts are present. The rifting was followed by a tilting of the shield of the Arabian Peninsula, lifting it to the northeast, toward the waters of the Arabian Gulf, and elevating Southern Yemen. The rift is accompanied by different tectonics than those of the folds and tilts of the neighboring sediment in Israel, Jordan, Lebanon, and Syria. Fig. 2.4.3 shows the post-Cretaceous structural movements in the Middle East.

The Great Rift Valley, stretching from Southwest Asia's northern Syria to East Africa's central Mozambique, is a large geographical and geological feature. Owing to the geological rifting cycle, it is a complex feature that involves many platelets of the lithosphere (Ponikarov, 1967; Mckenzie, 1970; and Abed, 1982). The rift valley is 30–100 km long and several

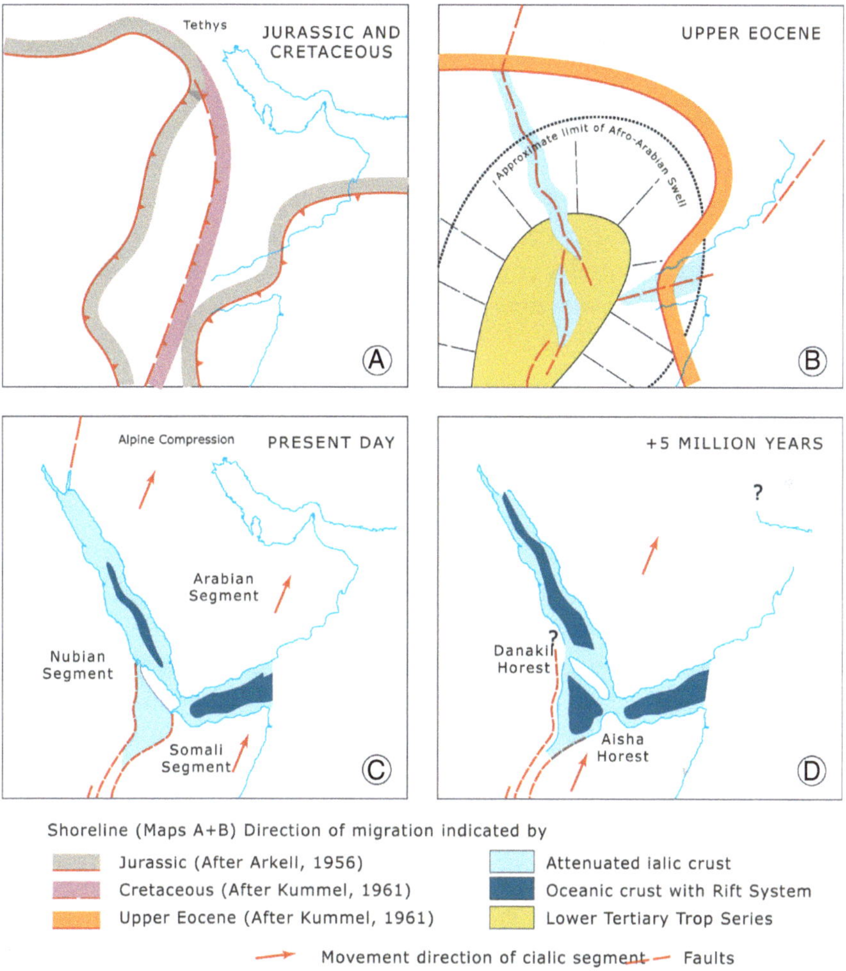

Fig. 2.4.3 ***The post-Cretaceous structural movements in the Middle East.***

hundred to several thousand meters deep. The major rift system extends from northern Lebanon to southern Mozambique. The most northerly portion of the Rift is Beqaa Valley in Lebanon bordering the "Lebanon Mountains and Anti-Lebanon range" (Walley, 1998; Quennell, 1956; Abed, 1982). Hula Valley, between the Mountains of Galilee and the Golan Heights, is located further south. The Jordan River rises here and flows into the Sea of Galilee in Israel, then southward through Lake Hula and then proceeds into "the Dead Sea on the Israeli–Jordanian border" (Fig. 2.4.2.) to the south across the Jordan Valley. The Wadi Arabah, the Aqaba Gulf, and the Red Sea (Cochran, 1983; Garfunkel et al., 1981) continue the rift south from the Dead Sea. "The southern end of the Red Sea" is a void in the rift. Ethiopia and Eritrea, the "Afar Triangle or Danakil Depression, are the probable places" of a three-piece intersection that may be underlined by a mantle plume. The Gulf of Aden is an eastern extension of the rift, which was connected to the Horn of Africa before the rift formed—and from that point the rift continues to form as part of the India Ocean's mid oceanic ridge. The Great Rift valley continues in a southwestern direction and splits the ancient Ethiopian mountain range in two parts.

Tertiary orogeny

Sadek (1977), Kent (1978), and Nehlig (1999) identified a tertiary orogeny in the MENA region. This can be best illustrated in the folding mountains that range from the Zagros to the Taurus Mountains and down to Oman. Tectonic styles vary widely. The Kyrenia range contains "*en echelon* folding," while in the "Palmyrean chains and the unusual Djebel Sin jar" are various subsidiary tectonic manifestations. The Syrian and Lebanese mountains are bound by rifts, but in Djebel Akhdar of Cyrenaica, folded mountain ranges return. Further to the West, deep sediments are found before being separated from the area with the entry of the Tunisian and Algerian Atlas and Anti-Atlas Fold.

The Zagros, which are distinguished by very long continuous anticlinal stripes, can reach 100 or 200 km without any splits, thanks to the extensive creation of a deep limestone floor. With the mountain chain, the soft rocks of the Fars Group have been eroded and are now limited to synclinal belts. Initially, the hard outer surfaces of the mountains were the immediately underlying Asmari Limestone, the most important historic reservoir, where the rock in the foothills was largely locked, but also bridged to reveal lower Bangestan calcareous stone and the numerous Lower Cretaceous and Jurassic calcareous stones. Most of these groups have a thickness of 500–1000 m and the boundary between the limestone eminences is argillaceous. Occasionally the Lower Paleozoic rocks are exposed to irregular amplitude folds (Kent, 1978; Ameen, 1991; Falcon, 1974; Jackson et al., 1981; Neugebauer, 1995). Kent has reported that to the northeast of the Zagros and extending to the Elburz mountains, the interior Iran includes several broad plains, broken locally by small areas of isolated hills, often consisting of limestone rocks, but sometimes of bedded igneous rocks. Mainly Eocene and Tertiary marls are a major part of the surface rocks; the hills are scattered with either Miocene limestone or Eocene Tufts and lava (which are locally oil bearing if they occur at depth, as they are at Qum). Nevertheless, the central plains are part of an inner drainage basin and the eastern parts of the region is a salt desert, in which soils are filled with gypsum and sodium chloride, which are also churned by their continuous recrystallization. Many of the mountains in the Dashte Kavir are themselves growing salt plugs.

Quaternary cover

Brice (1978), Al-Sayari and Zotel (1978), and Glennie (1978, 1997, 2002) address the Quaternary cover of the MENA region in detail. They show the large quartz sand volumes eroded from older sandstones and may demonstrates how the Paleozoic and other continental sandstones were deposited. Where sand is absent, the typical cover is gravel plains (sarir) with lag gravel, with varying degrees of stoniness and ruggedness. Many surfaces, primarily carbonates but occasionally sulfates, are cemented with calcrete. The lacustrine deposits of are always large in volume, produced in pluvial lakes. Dropping baseline rates allow terraces to demonstrate past erosion. Eolian erosion is common, as in Kharga in Egypt, in places that deflate to the water table. Saline mud flats or playas (sabkha) occur at inland drainage baselines; coastal sabkha is also of interest. On the outskirts of the mountains there are major fans and deltas as diminishing water flow drops the load of eroded material. Remnants of old drainage systems can be seen in many regions and remnants of pluvial conditions and even today occasional floods are recorded at very sporadic intervals.

Baltzer and Purser (1990) reported that the Persian Gulf reached as far northwest as Syria in the early Quaternary and that the deltaic network of the river Tigris and Euphrates have been thought to have filled at least half the initial basin and reduced the size of the Persian Gulf. The Levantine zone of Badia consists primarily of Paleogene to Quaternary continental alkaline—olivine basalts and tuffs, bordered by Paleogene, Cretan, Neogene, and Quaternary carbonates and clastics to the east and southwest. Due to recent clastic studies, many small, closely separated sandstone outcrops have been assigned to the Asraq formation within a rift region, bounded by the northwest–southeast trend Fuluq-Sirhan faults. Paleogene and Neogene rocks dominated the study area in southern Badia.

The oldest quaternary sediments in the Arabian Gulf and according to Glennie and Singhvi (2002) are alluvial fans, which spread away from the major mountains, for example, in Oman. In Sabkha Matti, in the Northwest part of the region, fluvial sequences pre-Holocene are also exposed (Goodall, 1995). Fluvial channels are still active intermittently, but erosion appears to be more dominant than accretion, partly due to deflation, particularly in the upper-fan regions. The sedimentary sequences in the area, which immediately underlie the sediments of the "desert," are not well known. Apart from the older deposits of alluvial fans, the Pliocene can only be interpreted by a phase of erosion (see, for example, Le Métour et al., 1995; Alsharhan and Nairn, 1997). The exposed rocks provide evidence of a transition from shallow sea carbonates to lacustrine deposits in southern Oman during the Miocene period (Jones and Racey, 1994). Baynunah Formation fluvial deposits (Friend, 1999) and the Shuweihat Formation (Bristow, 1999) underly eolian sequences in the western Emirates.

Glennie and Singhvi (2002) also reported that the Wahiba and Rub al Kali sands create a specific desert climate in terms of eolian processes, affected by two orthogonal wind systems that respond to different climate regimes. In comparison, eolian sediments are dominated by the winter Shamal winds from NNW to SSE and Summer Monsoon from South–West to North–East. As a result, the Emirates' eolian accretion has decreased due to its reliance on the water levels in the Arab Gulf, since there has been a glacially driven increase in world sea level. A comparative analysis of the sediment deposition timing in the region thus has the ability to recreate the amplitude and direction of past air circulation trends in the region. In order to understand the vital limit and response times of sedimentary cycle in this desert, this can be contrasted with the Arabian Sea records off Oman's coast.

Eventually, during the Pleistocene and the early Holocene, weather conditions contributed to the accumulation of massive water masses in the Sahara of Middle East and North Africa, these water masses formed playas and sabkhas, where calcrete is also found with other minerals that evaporated. However, along today's shorelines, physiochemical processes concentrated large amount of heavy or black sand as in Khorfacan in the Arabian peninsula, Dead Sea in Jordan, and along the Red Sea coasts and the sabkhas near Shalateen, as well as beneath sand west of Alexandria.

Geotectonic framework and mineralization

Based on this general geodynamic structure, different forms of mineralization found in the Middle East can be put into their structural and metallogenic background. In the Saudi portion of the Arabian Shield, mineralization is reported that includes ophiolite-related facies of Cr–Ti–Fe–Ni–Cu and some of the volcanic massive sulfide ("VMS") and Cu ore bodies. Other forms of mineralization are associated with island arc volcanic activity, such as epithermal ore bodies and VMS submarine deposits in submarine environments and arc-plutonic porphyry stocks. The rest of the types resulted from remobilization of pre-existing mineralization, especially mesothermal gold related to accretion of arcs and PanAfrican tectonics. Mineralization (tin–tungsten) at intrusive post-tectonic peraluminous plutons sites was also reported, as well as other forms of mineralization in sandstone across the Red Sea that includes Pb–Zn mineralization in carbonate platform or paleochannel environments. Other reported mineralization is related to paleoplacers (gold), lateritic paleosurface weathering rocks (bauxite), or gold-bearing VMS gossans (Nehlig et al., 1999).

Rare earths, thorium, and uranium mineralization occurrences have also been reported by Nehlig et al. (1999) in the Samran-Mahd Adh Dhahab-Jabal Sayid field. They suggested that the alkaline Nb–Y–Zr Jabal Hadb/Ash Sharar granite, south of Jabal Sayid, comprises 23 Mt of reserves at 0.13% Nb; 0.13% Ce, >1.7% Zr, and 134 ppm U (Elliott, 1997). Rare earth element (REE) is present in bastnaesite, doverite, monazite, and synchisite; Nb is associated with pyrochlore, Zr with zircon, and thorium to thorite and thorianite. The monzogranitic complex hosting these rocks has been dated to 573 Ma (Calvez and Kemp, 1982). The intrusion of quartz–syenite of Hamra in the northeast of Qalat Zummurud is one of the richest intrusions of REE known in KSA. The resource estimated in silexite with Fe oxides, bastnaesite, and monazite is around 18 Mt at 0.17% Nb, 0.26% La, 0.34% Ce, and 1.3% Zr. Other mineralization zones are found in (1) the Wadi Azlam area, situated in the south and hosts seven Cu–Zn volcano-sedimentary events, (2) banded iron formations and japillite around Wadi Sawawin, and (3) to the north, some incidents include young intrusives, either Cu–Mo type (porphyries) or peralkaline intrusive REE-Nb.

In summary, the key characteristics that are relevant to Chapter 5 of this book that define the MENA region can be attributed to the following causes:

1. Extensional tectonics in Eastern and Southern Arabia resulting from the expansion of seafloors in the Red Sea and the Gulf of Aden and with the formation of incipient seafloor spreading.
2. Compressional tectonic forces in the Taurus–Zagros–Oman Orogenic Region in the north and northeast, resulting from collision of continent–continent; and strike-slip faulting along the Levant Fracture Region or the Dead Sea Rift.

3. The transgression and regression of the Tethys Ocean.
4. The Nubo-Arabian Shield and associated mechanical and chemical weathering products, which were transported into the shelf of the Tethys Ocean.

More information on the geology of MENA region can be found in Beydoun (1977), Brueckner et al. (1987), Dubertret (1932), Dubertret (1955), Duncan et al. (1990), Falcon (1961), Garfunkel and Bartov (1977), Girdler (1982), Girdler (1990), Girdler and Southren (1987), Kelling et al. (1987), Kesten et al. (2006), Alsharhan and Kendall (1986), Koop and Stoneley (1982), Quennell (1958), Garfunkel (1981), Garfunkel (1997), and Bahat and Rabinovitch (1983).

References

Abed, A.M., 1982. Geology of Jordan. Al Nahda Al Isla miah Library, Amman.

Abed, A.M., 1985. On the supposed Precambrian palaeosuture along the Dead Sea Rift, Jordan. J. Geol. Soc. Lond. 142, 527–531.

Abed, A.M., 1988. Precambrian paleosuture along the dead-sea rift-reply. J. Geol. Soc. 145, 516–517.

Abed, A.M., Sadaqah, R., 1998. Role of Upper Cretaceous oyster bioherms in the deposition and accumulation of high-grade phosphorites in central Jordan. J. Sediment. Res. 68 (5), 1009–1020.

Agar, R.A., 1987. The Najd fault system revisited; a two-way strike-slip orogen in the Saudi Arabian Shield. J. Struct. Geol. 9 (1), 41–48.

Al-Sayari, S.S., Zötl, J.G., 1978. Quaternary Period in Saudi Arabia, 1: Sedimentological, Hy-drogeological, Hydrochemical, Geomorphological and Climatological Investigations in Centraland Eastern Saudi Arabia. Springer, New York.

Alsharhan, A.B., Kendall, C.G.S.C., 1986. Precambrian to Jurassic rocks of Arabian Gulf and adjacent areas: their facies, depositional setting, and hydrocarbon habitat. Am. Assoc. Petrol. Geol. Bull. 70, 977–1002.

Al-Sharhan, A.S., Kendall, C.G.S.C., 1986. Precambrian to Jurassic rocks ofArabian Gulf and adjacent areas: their facies, depositional setting,and hydrocarbon habitat. Am. Assoc. Petroleum Geol Bull 70, 977–1002.

Alsharhan, A.S., Nairn, A.E.M., 1997. Sedimentary Basins and Petroleum Geology of the Middle East1997. Elsevier, Amsterdam, p. 843 with an Appendix of oil and gas field data (p. 97), ISBN 0-444-82465-0.

Altherr, R., Henjes-Kunst, F., Baumann, A., 1990. Asthenosphere versus lithosphere as possible sources for basaltic magmas erupted during formation of the Red Sea: constraints from Sr. Pb and Nd isotopes. Earth Planet. Sci. Lett. 96, 269–286.

Ameen, M.S., 1991. Possible forced folding in the Taurus-Zagros belt of Northern Iraq. Geol. Mag. 128 (6), 561–584.

Andrew, G., 1948. Geology of the Sudan. Chapter IV in Agriculture in the Sudan. Tothill, 84–128.

Attia, M.I., 1955. Topography, geology and iron-ore deposits of the district east of Aswan. Geol. Survey publications, cairo, 99–114.

Bahat, D., Rabinovitch, A., 1983. The initiation of the Dead Sea rift. J. Geol. 91, 317–332.

Baker, V.R., 1996. The Pragmatic Roots of American Quaternary Geology and Geomorphology. Geomorphology 16, 197–215.

Baker, V.R., 1998. Catastrophism and Uniformitarianism: Logical Roots and Current Relevance in Geology. In: Blundell, D.J., Scott, A.C. (Eds.). Lyell: The Past is the Key to the Present. Geological Society of London Special Publications, 143, pp. 171–182.

Baltzer, F., Purser, B., 1990. Modern alluvial fan and deltaic sedimentation in a foreland tectonic setting: the Lower Mesopotamian Plain and Arabian Gulf. Sediment. Geol. 67, 175–197.

Bellen, R.C. van, H.V. Dunnington, R. Wetzel, and D.M. Morton, eds., 1959. Lexique stratigraphic international: Paris, v. III, Asie, Fascicule 10a Iraq, p. 333.

Ben-Avraham, Z., 1978. The structure and tectonic setting of the Levant continental margin, eastern Mediterranean. Tectonophysics 46, 313–331.

Bender, D.F., 1968. Geologie von Jordanien, Beiträge zur regionalen Geologie der Erde. Gebrüder Borntraeger, Berlin, pp. 187–190.

Bentor, Y. K., and Vroman, A., 1960. The Geological Map of Israel on 1: 100,000 scale, Ser. l, Sheet l6, Mt. Sdom. Geol. Surv. of Israel.

Best, J.A., Barazangi, M., Al-Saad, D., Sawaf, T., Gebran, A., 1993. Continental margin evolution of the northern Arabian platform in Syria. Am. Assoc. Petrol. Geol. Bull. 77 (2), 173–193.

Beydoun, Z.R., 1964. The stratigraphy and structure of the eastern Aden Protectorate. HMSO, London.

Beydoun, Z., 1977. The Levantine countries: the geology of Syria and Lebanon (maritime regions). In: Nairn, A.E.M., Kanes, W.H., Stehli, F.G. (Eds.), The Ocean Basins and Margins, The Eastern Mediterranean. Plenum Press, New York, London, pp. 319–353.

Beydoun, Z.R., 1992. Petroleum in the Zagros Basin: a Late Tertiary Foreland Basin overprinted onto the outer edge of a vast hydrocarbon-rich Paleozoic-Mesozoic passive-margin shelf: Chapter 11. In: Leckie, R.M.a.D. (Ed.), Foreland Basins and Foldbelts. American Association of Petroleum Geologists Memoir, p. 55.

Bouabdellah, M., Slack, J.F., 2016. Geologic and metallogenic framework of North Africa. In: Bouabdellah, M., Slack, J.F. (Eds.), Mineral Deposits of North Africa. Mineral Resource Reviews. Springer International Publishing, Berlin. doi:10.1007/978-3-319-31733-5 Springer-VerlagISBN: 978-3-319-31731-1.

Brice, W.C., 1978. The environmental history of the Near and Middle East since the Last Ice Age. Academic Press, London.

Brinkmann, R., 1971. The Geology of western Anatolia. Geology and History of Turkeyedited by : A. S. campbell, Tipoli, pp. 171–190.

Bristow, C.S., 1999. Aeolian and Sabkha Sediments. Fossil Vertebrates of Arabia, 50.

Brueckner, H.K., Zindler, A., Seyler, M., Bonatti, E., 1987. Zabargad and the Pan-African and Miocene isotopic evolution of the sub-Red Sea mantle and crust. Abstr Programs – Geol. Soc. Am. 19 (7), 603.

Burgoynea, P.M., van Wykb, A.E., Andersona, J.M., Schrirec, B.D., 2005. Phanerozoic evolution of plants on the African plate. J. Afr. Earth Sci. 43 (1-3), 13–52.

Burke, K., Gunnell, Y., 2008. The African erosion surface: a continental-scale synthesis of geomorphology, tectonics, and environmental change over the past 180 million years. In: Burke, K., Gunnell, Y. (Eds.), Continental-Scale Synthesis of Geomorphology, Tectonics, and Environmental Change over the Past 180 Million Years. Geological Society of America, Memoirs, 201, pp. 1–66. https://doi.org/10.1130/2008.1201.

Burollet, P.F., 1960. Lexique Stratigraphique Internal, Vol IV Afrique, No IV a Libye. Centre National de Recherche Sciences, Paris, p. 62.

Calvez, J. Y., Kemp, J., 1982. Geochronological Investigations in the Mahd Adh Dhahab Quadrangle Saudi Arabia: Deputy Ministry for Mineral Resources Technical Record BRGM-TR-02-5, p. 41.

Camp, V.E., Hooper, P.R., Roobol, M.J., White, D.L., 1987. The Madinah eruptin, Saudi Arabia: magma mixing and simultaneous extrusion of three basaltic chemical types. Bull. Vocanol. 49, 489–508.

Camp, V.E., Roobol, M.J., 1992. Upwelling asthenosphere beneath western Arabia and its regional implications. J. Geophys. Res. 97, 15,255–15,271.
Chazot, G., Bertrand, H., 1993. Mantle sources and magma-continental crust interactions during early Red Sea-Aden rifting in Southern Yemen: elemental and Sr, Nd, Pb isotope evidence. J. Geophys. Res. 98, 1819–1838.
Cochran, J.R., 1983. Model for the development of the Red Sea. Am. Assoc. Petrol. Geol. Bull. 67, 41–69.
Crelling, J.C., Thomas, K.Mark, 1994. Review of some recent research on the combustion properties of coal macerals: Preprints - Div. of Fuel Chem. Am. Chem. Soc. 39 (1), 209–213.
De Sitter, L., 1962. Structural development of the Arabian Shield in Palestine. Geol. Mijnbouw 41, 116–124.
Delfour, J., 1970. Le groupe J'Balah, une nouvelle unité du bouclier Arabe. Bureau de Recherches Géologiques et Minières, Bulletin 4 (2), 19–32.
Dubertret, L., 1932. Les formes structurales de la Syrie et de la Palestine. C. R. Acad. Sci. Colon. 195, 66.
Dubertret, L., 1955. Carte Geologique du Liban. Ministere des Travaux Publics.
Dubertret, L., 1970. Review of structural geology of the Red Sea and surrounding areas. Philosophical Transactions for the Royal Society of London. Series A, Mathematical and Physical Sciences, 9–20.
Dubray, E.A., Stoeser, D.B., Mckee, E.H., 1991. Age and petrology of the Tertiary As-Sarat Volcanic Field, Southwestern Saudi-Arabia. Tectonophysics 198 (2-4), 155–180.
Duncan, I.J., Rivard, B., Arvidson, R.E., Sultan, M., 1990. Structural interpretation and tectonic evolution of a part of the Najd Shear Zone (Saudi-Arabia) using landsat thematic-mapper data. Tectonophysics 178 (2-4), 309–335.
Falcon, N.L., 1961. Major earth-flexing in the Zagros Mountains of southwest Iran. Quart. J. Geol. Soc. Lond. 117 (4), 367–376.
Falcon, N.L., 1974. Southern Iran: Zagros Mountains. In: Spencer, A.M. (Ed.), Mesozoic-Cenozoic Orogenic Belts, Data for Orogenic Studies. Geological Society of London, London, pp. 199–211.
Fisher, R.V., 1961. Proposed classification of volcaniclastic sediments and rocks. Geol. Soc. Am. Bull. 72, 1409–1414.
Flügel, H.W., 1964. The Geology of the Upper Djadjerud and Lar Valleys (N-Iran). Rivista Italiana di Paleontologia e di Stratigrafia 70, 403–444 pls. 28–34.
Freund, R., 1965. A model of the structural development of Israel and adjacent areas since Upper Cretaceous times. Geol. Mag. 102 (3), 189–205.
Friend, C.R.L., 1999. When Continents Collide: Geodynamics and Geochemistry of Ultrahigh-Pressure Rocks. Mineralogical Magazine.
Garfunkel, Z., 1981. Internal structure of the Dead Sea leaky transform (rift) in relation to plate kinematics. Tectonophysics 80, 81–108.
Garfunkel, Z., 1997. The history and formation of the Dead Sea basin. In: Niemi, T.M., Avraham, Z.B., Gat, J.R. (Eds.), The Dead Sea, The Lake and Its Setting, Oxford Monographs on Geology and Geophysics, pp. 36–56.
Garfunkel, Z., Bartov, Y., 1977. The tectonics of the Suez Rift. Israel Geol. Surv. Bull. 71, 44.
Garfunkel, Z., Zak, I., Freund, R., 1981. Active faulting in the Dead Sea Rift. Tectonophysics 80, 1–26.
Garnfunkel, Z., 2000. History and paleogeography during the Pan-African orogen to stable platform transition: Reappraisal of the evidence from the Elat area and the northern Arabian-Nubian shield. Israel Journal of Earth Sciences 48, 135–157.
Gass, I.G., Gibson, I.L., 1969. Structural evolution of the rift zones in the Middle East. Nature 221 (5184), 926–930.
Girdler, R.W., 1982. The importance of the Jordanian Rift to studies of the Red Sea and Gulf of Aden. Proc. First Jord. Geol. Conf. 1, 503–522.

Girdler, R.W., 1990. The Dead Sea transform fault system. Tectonophysics 180, 1–13.
Girdler, R.W., Southren, T.C., 1987. Structure and evolution of the northern Red Sea. Nature 330, 716–721.
Glennie, K.W., 1997. Quaternary Arabia: Is it a palaeogeographic and climatic analogue of NW Europe's Permian Upper Rotliegend 2? J. Pet. Geol. 20 (1), 100–104.
Glennie, K.W., Mudd, G.C., Nagtegaal, P.J.C., 1978. Depositional environment and diagenesis of Permian Rotliegendes sandstones in Leman Bank and Sole Pit areas of the UK southern North Sea. J. Geol. Soc. London 135, 25–34.
Glennie, K.W., Singhvi, A.K., 2002. Event stratigraphy, paleoenvironment and chronology of SE Arabian deserts, Quat. Sci. Rev. 21 (7), 853–869. doi:10.1016/S0277-3791(01)00133-0.
Goodall, P., Johnson, S.G., Wood, E., 1995. Laser ablation inductively coupled plasma atomic emission spectrometry of a uranium-zirconium alloy: Ablation properties and analytical behavior. Spectro Acta Part B 50, 1823–1835.
Haberland, C., Maercklin, N., Kesten, D., Ryberg, T., Janssen, C., Agnon, A., Weber, M., Schulze, A., Qabbani, I., El-Kelani, R, 2007. Shallow architecture of the Wadi Araba fault (Dead Sea Transform) from high-resolution seismic investigations. Tectonophysics 432, 37–50. doi:10.1016/j.tecto.2006.12.006.
Hirsch, F., Picard, L., 1988. The Jurassic facies in the Levant. J. Petrol. Geol. 11, 277–308.
Howari, F., Goodell, P., Salman, A., 2016. Metallogenic evolution of uranium deposits in the Middle East and North Africa deposits. J. Afr. Earth Sci. 114, 30–42.
Husseini, M., 1988. The Arabian Infracambrian extensional system. Tectonophysics 148, 93–103.
Immenhauser, A., 2005. High-rate sea-level change during the Mesozoic: new approaches to an old problem. Sediment. Geol. 175 (1-4), 277–296.
Jackson, J.A., Fitch, T., McKenzie, D.P., 1981. In: McClay, K., Price, N. (Eds.), Active thrusting and the evolution of the Zagros fold belt. Geological Society, London, Special Publications, 9 (1), pp. 371–379.
Johnson, P.R., Woldehaimanot, B., 2003. Development of the Arabian-Nubian Shield: perspectives on accretion and deformation in the northern East African Orogen and the assembly of Gondwana. Geological Society. London, Special Publications, 206 (1), pp. 289–325.
Jones, R.W., Racey, A., 1994. Cenozoic stratigraphy of the Arabian Peninsula and GulfSimmons MD, Micropalaeontology and Hydrocarbon Exploration in the Middle East. Chapman and Hall, London, pp. 273–307.
Kamen-Kaye, M., 1971. A review of depositional history and geological structure in Turkey, Geology and History of Turkey. Petroleum Exploration Society of Libya, 13th Annual Field Conference, 111–137.
Kelling, G., Gokcen, S., Floyd, P., Gokcen, N., 1987. Neogene tectonics and plate convergence in the eastern Mediterranean: new data from southern Turkey. Geology 15, 425–429.
Kent, D.M., 1969. The geology of the Upper Devonian Saskatchewan Group and equivalent rocks in western Saskatchewan and adjacent areas. Sedimentary Geology Division, Geological Sciences Branch, Department of Mineral Resources.
Kent, P.E., 1978. Middle East—the geological background. Q. J. Eng. Geol. Hydrogeol. 11, 1–7.
Kesten, D., 2004. Structural Observations at the Southern Dead Sea Transform from Seismic Reflection Data and ASTER Satellite Images. Ph.D. thesis. University of Potsdam, Germany.
Kesten, D., Weber, M.H., Haberland, C., Janssen, C., Agnon, A., Bartov, Y.DESERT Group, 2006. Combining satellite and seismic images to analyze the shallow structure of the Dead Sea Transform near the DESERT transect. Int. J. Earth Sci.
King's College, LondonTaylor Dr., J.H., 1949. Nature 163, 474–475.

Kolodny, Y., 1967. Lithostratigraphy of the Mishash formation, northern Negev. Israel Journal of Earth Sciences 16 (2), 57–73.
Koop, W.J., Stoneley, R., 1982. Subsidence history of the Middle East Zagros basin, Permian to recent. Phil. Trans. R. Soc. Lond. 305, 149–168.
Kroner, A., Williams, I.S., Compston, W., Baur, N., Vitange, P.W., Perara, L.R.K., 1987. Zircon ion microprobe dating of high-grade rocks in Sri Lanka. J. Geol. 95, 775–791.
Le Métour, J., Michel, J.C., Béchennec, F., Platel, J.-P., Roger, J, 1995. Geology and Mineral Wealth of the Sultanate of Oman. Muscat, Oman: Ministry of Petroleum and Minerals, Directorate General of Minerals, Sultanate of Oman. Muscat and Bureau de Recherches Géologiques et Minières, France, p. 285.
Lovelock, P.E.R., 1984. A review of the tectonics of the northern Middle East region. Geol. Mag. 121, 577–587.
McBride, J.H., Barazangi, M., Best‡, J., Al-Saad, D., Sawaf, T., Al-Otri, M., Gebran, A., 1990. Seismic reflection structure of the intracratonic Palmyride fold-thrust belt and surrounding Arabian platform, Syria. Am. Assoc. Pet. Geol. Bull. 74, 238–259.
McKee, E.D., 1963. Origin of the Nubian and similar sandstones. Geol. Rundsch. 52 (2), 551–587.
McKee, E.D., Sterrett, T.S., 1961. Laboratory Experiments on Form and Structure of Longshore Bars and Beaches. In: Peterson, J.A., Osmond, J.C. (Eds.), Geometry of Sandstone Bodies. American Association of Petroleum Geologists, pp. 13–28.
McKee, J.E., Wolf, H.W., 1963. Water Quality Criteria. California State Water Quality Control Board Publication, pp. 548 3-A.
McKenzie, D., Bickle, M.J., 1988. The volume and composition of melt generated by extension of the lithosphere. J. Petrol. 29, 625–679.
McKenzie, D., Davies, D., Molnar, P., 1970. Plate tectonics of the Red Sea and East Africa. Nature 226, 243–248.
McKenzie, D.P., 1970. Plate tectonics of the Mediterranean Region. Nature 226, 239–243.
Mouty, M., Delaloye, M., Fontignie, D., Piskin, O., Wagner, J.-J., 1992. The volcanic activity in Syria and Lebanon between Jurassic and Actual. Schweiz. Mineral. Petrogr. Mitt. 72, 91–105.
Neev, D., 1975. Tectonic evolution of the Middle East and the Levantine basin (easternmost Mediterranean). Geology 3 (12), 683–686.
Nehlig, P., Salpeteur, I., Asfirane, F., Bouchot, V., Eberlé, J.M., Genna, A., Kluyver, H.M., Lasserre, J.L., Nicol, N., Recoche, G., Shanti, M., Thiéblemont, D., Tourliere, B., 1999. The mineral potential of the Arabianshield: a reassessment, Proceedings of the IUGS/UNESCO Meeting on the "Base and Precious Metal Deposits in the Arabian Shield". Jeddah, November, p. 12e19.
Neugebauer, J., 1995. Structures and kinematics of the North Anatolian Fault Zone, Adapazan-Bolu Region, Northwest Turkey. Tectonophysics 243 (1-2), 119–134.
Pik, R., Deniel, C., Coulon, C., Yirgu, G., Marty, B, 1999. Isotopic and trace element signatures of Ethiopian fl ood basalts: Evidence for plume-lithosphere interactions. Geochim. Cosmochim. Acta 63, 2263–2279. doi:10.1016/S0016-7037(99)00141-6.
Ponikarov, V.P., Kazmin, V.G., Mikhailov, I.A., Razvaliayev, A.V., Krasheninnikov, V.A., Kozlov, V.V., … & Mirzayev, K.M., 1967. The geology of Syria. Explanatory notes on the geological map of Syria, scale, 1(500,000).
Powers, R.W., Ramirez, L.F., Redmon, C.D., Elber Jr., E.L., 1966. Geology of the ArabianPeninsula: sedimentary geology of Saudi Arabia. US Geol. Surv. Prof. Pap. 560–D, 147.
Pruvest, P., 1951. L'Infracambrien. - Bull. Soc. Belg. Géol. Pal. Hydr. 60 (1), 43.
Pufahla, P.K., Grimma, K.A., Abedb, A.M., Sadaqahb, R.M.Y., 2003. Upper Cretaceous (Campanian) phosphorites in Jordan: implications for the formation of a south Tethyan phosphorite giant. Sedimentary Geol. 161 (3-4), 175–205.

Quennell, A.M., 1956. Tectonics of the Dead Sea Rift, Proceedings of the 20th IGC in Mexico, 385–403.
Quennell, A.M., 1958. The structural and geomorphic evolution of the Dead Sea Rift. Quart. J. Geol. Soc. Lond. 114, 1–24.
Quennell, A.M., 1984. The Western Arabia rift system. In: Dixon, J.E., Robertson, A.H.F. (Eds.), The Geological Evolution of the Eastern Mediterranean. Blackwell Scientific Publications, Oxford, pp. 775–788.
Reiss, Z., 1988. Assemblages from a Senonian high-productivity sea. Revue de Paléobiologie, 323–332.
Sadek, A., 1977. Early paleozoic sediments of the Zagros-Taurus-ranges. Geol. Rundsch. 66, 263–276.
Sage, L., Letouzey, J., 1990. Convergence of the African and Eurasian plate in the Eastern Mediterranean. In: Letouzey, J. (Ed.), Petroleum and Tectonics in Mobile Belts; Proceedings of the IFP Exploration and Production Research Conference. Editions Technip, Paris, pp. 49–68.
Said, R., 1962. The Geology of Egypt. Elsevier, Amsterdam, New York, p. 377.
Sanford, B.V., Thompson, F.J., McFall, G.H., 1985. Plate tectonics – A possible controlling mechanism in the development of hydrocarbon traps in southwestern Ontario. Bull. Can. Petrol. Geol. 33 (1), 52–71.
Sharland, P.R., Archer, R., Casey, D.M., Davies, R.B., Hall, S.H., Heward, A.P., Horbury, A.D., Simmons, M.D., 2001. Arabian Plate Sequence Stratigraphy. GeoArabia Special Publication 2, 371.
Shata, A.A., 1955. An introductory note on the geology of northern portion of the Western Desert of Egypt. Bull. Inst. Desert d' Egypte T V (1), 96–106.
Shaw, J.E., Baker, J.A., Menzies, M.A., Thirlwall, M.F., Irbrahim, K.M., 2003. Petrogenesis of the largest intraplate volcanic field on the Arabian Plate (Jordan): a mixed lithosphere–asthenosphere source activated by lithospheric extension. J. Petrol. 44 (9), 1657–1679.
Shukri, N.M., Said, R., 1944. Contribution to the geology of the Nubian Sandstone, 1. Field observation and mechanical analysis. Foad I Univ. Cario Bull., Fac. Sci. 25, 149–172.
Stein, M., Hofmann, A.W., 1992. Fossil plume head beneath the Arabian lithosphere? Earth Planet. Sci. Lett. 114 (1), 193–209.
Stewart, K., Rogers, N., 1996. Mantle plume and lithosphere contributions to basalts from southern Ethiopia. Earth Planet. Sci. Lett. 139, 195–211.
Stoeser, D.B., Camp, V.E., 1985. Pan-African microplate accretion of the Arabian Shield. Geol. Soc. Am. Bull. 96, 817–826.
Turkünal, S., 1953. Géologie de la région de Hakkâri et de Baskale Turquie: Hakkâri ve Baskale, Bölgelerinin Jeolojisi (Doctoral dissertation, Enstitüsü Matbaasi).
Walley, C.D., 1998. Some outstanding issues in the geology of Lebanon and their importance in the tectonic evolution of the Levantine region. Tectonophysics 298 (1-3), 37–62.

CHAPTER 3

Exploration of uranium with case studies

CHAPTER 3.1

Introduction

Uranium and other radioactive minerals are widely spread in many types of crustal rocks. Their distribution in various concentrations is subject to various factors making its exploration a relatively difficult task. The techniques used for surveying in uranium exploration and prospecting can be split in a traditional geological survey and radiometric survey. Uranium prospecting is little different than other forms of mineral exploration because some advanced instruments for detecting radioactive isotopes are needed such as advanced geophysical techniques, drilling equipment, and other tools to accomplish the uranium survey objectives.

However, any uranium survey should focus on the following points as will be described in the subsequent sections.

1. Rapid determination of unexplored (and poorly explored) areas that have potential uranium ore deposits.
2. Preliminary estimation of the uranium reserves that might be present in the target area.
3. Reveal the characteristic features of the metallic deposits and use them as a guide to predict new deposits.
4. Locate new types of uranium ore deposits, especially in metallic environments
5. Reduce the costs and time of exploration to move to the production stage.

National geological surveys, mining companies, and research institutions are now conducting uranium exploration and related geological interpretations using satellite imagery, spectrometer development and design, laser fluorometers for trace uranium analysis, mineral inclusion analyzers, radon scintillation analyzers, scintillation spectrometers, laser analyzers for trace substances, and high- and low-frequency dielectric separators. The systematic prospecting for uranium ores and their reserves calculation should pass through three main stages as described in Fig. 3.1.1.

Uranium Geology of the Middle East and North Africa
DOI: https://doi.org/10.1016/B978-0-323-90992-1.00003-8

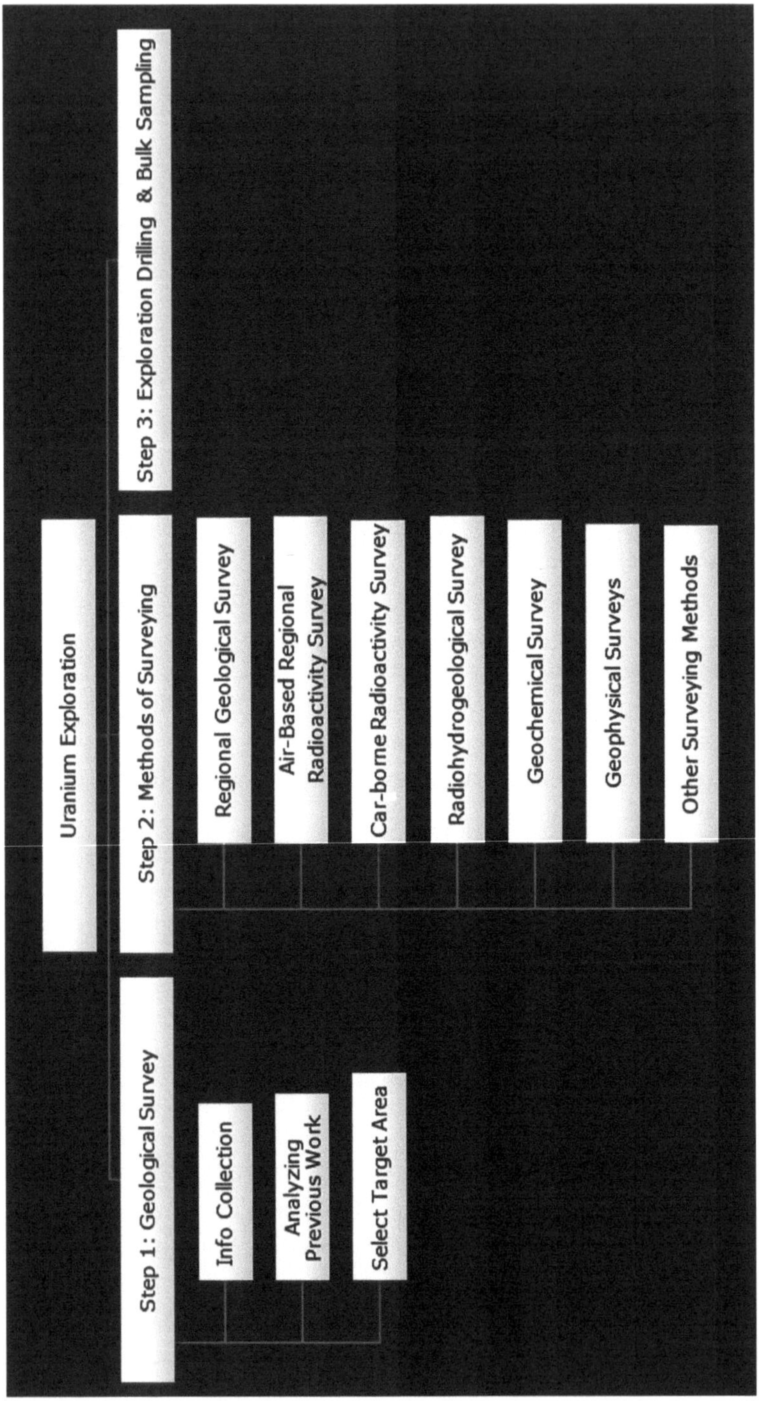

Fig. 3.1.1 *General consecutive steps in uranium exploration.*

CHAPTER 3.2

Geological survey

Different geological approaches are usually used to identify and map information about uranium ores exploration. The first phase in most development programs is to scope the area of interest that possibly host uranium ore deposits in economic concentrations. This initial step is often called prospecting. Prospecting results in the development of a geological map that forms a major reference source for other costly techniques. Melkov (1956) believes that the aim of a geological survey is not only to discern favorable districts but also to establish their relative potential to determine the sequence of recognition. The general principles, methods, and means of conducting a geological survey are well known and consist of the following

Information collection

It is the initial stage where uranium geologists customize a framework to organize and retrieve data so it can be used to determine the precise information of a certain region. In the past, before the advancement in the geographic information system, the easiest way to do that was to superimpose several transparent maps of different themes over one another. Maps with a scale of 1:1000 are often suitable for the collection of the required data that are an essential part of the exploration work. The themes include topographic, geological, tectonic, magnetic or radiation geophysical, and metallogenic maps covering the target area along with aerial photographs or satellite images.

Today, the integration of these data layers is time- and cost-effective using geographic information system. However, layering or superimposing these layers of information, will help the researchers to observe the relations and interpret the associations of such data to identify new targets for exploration. It is evident that collecting as much information as possible and analyzing them plays an important role in the selection of areas with high potential prospects of uranium ore deposits (Fig. 3.2.1).

At the beginning of the uranium exploration, several radioactive deposits with various radioactivity levels are determined. This is followed by a screening process to decide upon the explored uranium deposits, either to further explore them or dismiss them temporarily or permanently. However, it is necessary to maintain a record explaining the description of all conditions associated with each anomaly detected by the radiation

Fig. 3.2.1 ***Autunite, a secondary uranium mineral named after Autun in France.***

properties. Therefore, a standard sheet is prepared to report the characteristics of each anomaly. Then, these data are archived in a computer database to enable easy retrieval and usage of the obtained data.

Based on the radioactive anomalies, uranium geologists classify granite with 1500 count per second as granite with anomalous radiation. The rest of the geological formations or materials should have radiometric readings up to three times the normal reading (background) to be considered as a formation with abnormal radiation. After documenting the information and index-associated attributes of each radioactive anomaly in each site, uranium geologists prepare maps with a scale of 1:1000 borne by those sites. These maps are converted to a digital map and transferred to a proper database.

Select target area

When selecting a new region to be a target for uranium exploration, uranium geologists must first define or characterize the rocks that can be a source of uranium. Thus, adequate knowledge of plutonic and volcanic rocks is necessary because these rocks are an important source for many of the known uranium deposits located in various geological environments. It is known that acidic granite rocks usually have a good chance of hosting uranium deposits, and therefore considered a good potential source of uranium. Volcanic rocks could also be a source of uranium, especially rhyolite, peraluminous rocks, metaluminous tuff, and peralkaline tuff. These rocks often contain uranium in their glassy or very fine crystal groundmass and are excellent source rocks for uranium deposits that form through secondary processes associated with or following their crystallization. These processes include supergene, diagenetic, hydrothermal, or metamorphic events. To separate the uranium from the glassy or very fine crystalline groundmass, the rocks or the groundmass must be exposed to an alteration process that usually occurs either as a result of a temperature increase or decrease. Such temperature variations are associated with solutions or gases that involve re-crystallization aspects in the late stage of magma crystallization, granophyric crystallization, or fumerolic alteration. The decrease in temperature changes the rock texture and is often due to the impact of ground water through several stages during the weathering and diagenesis processes.

It is also noteworthy to mention that an adequate knowledge of the nature and properties of uranium found in plutonic and volcanic rocks could facilitate the process of selecting the target area. Thus, to identify good prospects, it is necessary to prepare additional maps such as the following:

- Map showing the distribution and composition of igneous rocks.
- Map illustrating the distribution and petrographic properties of plutonic rocks.
- Map displaying the age and petrographic properties of volcanic rocks in relation with the local stratigraphic unit in the area of interest.
- Map describing the known sites of uranium, thorium, rare earth elements, lithium, molybdenum, tin, fluorite, black quartz, kaolinite, felsic pegmatite, and boron.
- As most of the known uranium deposits are associated with acidic igneous rocks, the superimposition of previous maps over one another will help to identify areas with good prospects.

These areas can be arranged in the following order of decreasing potentiality:

- Areas near the basement rocks containing granitic rocks.
- Areas of volcanic and metamorphic rocks that have a volcanic origin.
- Areas of multi-type plutonic rocks.
- Areas containing plutonic rocks with acidic dykes or sills.

Melkov (1956) considers that based on a study of the geological structure of the target area and an analysis of the history and principal stages of its geological development, the area can be divided into different zones that are identical in their geological structure and history (tectogenesis, magmatism, sedimentation, paleography, etc.). Thereafter, it becomes possible to classify the regions whose geological history comprises individual phases, during which a combination of favorable conditions for the formation of some types of mineral deposits is met. In a uranium-prospecting survey, uranium geologists make a distinction for territories that may contain magmatic deposits and basins in which sedimentary and sedimentary metamorphic deposits may occur.

The most important radiological methods in the exploration of radioactive mineral deposits use instruments that can sense gamma radiation emanating from the rock formations bearing radioactive minerals. The use of these methods in the investigation of the atmosphere and the earth's surface is done using special devices such as Flash meters (scintillation counters), Geiger counters, and other spectroscopic-based methods. Also, meters emissivity (emanometers) are used for measuring radon gas and emitted gaps and spaces in rocks and soil. The limitation of radiological surveys is mostly due to the thick rock cover or soil layers. Therefore, subsurface surveying, such as gamma-ray well logging in drilled areas, is needed to determine the belts characterized by high levels of radioactivity.

The regional tools to explore for uranium depend mainly on airborne sensors. G.C Ridland, a geophysicist working at Port Radium in 1943, initially suggested using airborne detectors to search for radioactive minerals. In 1947, Eldorado Mining and Refining Limited (a Canadian Crown Corporation since sold to become Cameco Corporation) conducted the earliest reported trial of airborne radiation detectors (ionization chambers and Geiger counters). In 1949, Professors Pringle, Roulston, and Brownell of the University of Manitoba filed the first patent for a portable gamma ray spectrometer, the same year that they tested the first portable scintillation counter on the ground and in the air in northern Saskatchewan (Geological Survey of Canada, 2008).

The authors cite in different chapters of this book several reports, masters and doctoral dissertations that include a wealth of information on the Middle East and could help in the selection of the region with good prospects for the existence of uranium deposits. These reports focus on the geochemical composition of volcanic and plutonic rocks with an emphasis on (1) peraluminous granite characterized by low calcium, (2) metaluminous granite characterized by a high proportion of calcium, (3) peralkaline granite. The geochemical characteristics of these rocks greatly help in determining their potentiality as a source for uranium.

When identifying the favorable portions of the basins, sedimentary and sedimentary metamorphic uranium deposits must be considered in the

Proterozoic, Paleozoic (especially Ordovician, Carboniferous, and Permian), Mesozoic (Triassic, Jurassic, and Cretaceous), and Cenozoic (especially Paleogene) periods. In addition, it is important to consider the possibility of finding sedimentary uranium ores in Continental facies (such as lagoonal and lagoonal-lacustrine), especially in coal and sandstone. They are also found in littoral shallow-water transgressive facies, especially in sandstone, oolitic limestone, phosphorites, and clays rich in organic matter and elements such as phosphorus, vanadium, sulfur, fluorine, iron, molybdenum, barium, strontium, rare earth elements, and less frequently nickel, cobalt, zinc, lead, chromium, zirconium, hafnium, and scandium (Melkov, 1956).

Melkov (1956) emphasized that, in assessing particular regions and basins, it is important to consider not only the favorable conditions for the formation of commercial ores, but also the subsequent processes that could lead to the alteration or destruction of these deposits. Therefore, we first estimate the potential extent of the region's mineral degradation and deposits in the basin. The most desirable districts are those suffering from poor erosion where the uraniferous deposits were not lost. For sedimentary strata, we also find the folding characters and the degree of metamorphism that can lead to changes in stratification, simple regrouping and redistribution of uranium, and important changes in host rocks (such as re-crystallization and shale formation) and their properties.

CHAPTER 3.3

Geological basis of prospecting methods

There is currently interest in three types of uranium deposits: magmatic (mainly hydrothermal), sedimentary, and metamorphic sedimentary deposits. There is still inadequate analysis of their composition and distribution, although the characteristics of the favorable districts for each type and the structural features of the deposits have already been calculated on the basis of available data. It allows the geological splitting of the target region to coordinate the hunt for uranium deposits and greatly encourages identification within the appropriate districts (Melkov, 1956).

The search for commercial deposits in suitable districts and basins takes place over a variety of stages that vary in the amount of research involved and in the peculiarities of the methods and different techniques involved. The major peculiarity of uranium exploration is the widespread use of various radiometric methods. These methods are based on the radioactive nature of uranium, which naturally disintegrates and then turns into a sequence of other radioactive elements. Charged alpha and beta particles as well as gamma radiation are released when uranium is naturally disintegrated. Another uranium element, the luminescence of some of its uranyl compounds under ultraviolet light, is often used as an additional method (Melkov, 1956).

In major focus areas, uranium exploration typically starts with air-radiometric, car-borne-radiometric, and radio hydrogeological detection identifying regions of radioactive anomalies, that is, elevated (compared to the surrounding area) gamma radiation regions and concentration of radioactive elements in subterranean waters (Melkov, 1956). This method of prospecting is focused on the ability of uranium and its daughter to form radioactive elements (ionium, radium, and radon) around the upper oxidized elements of deposits and outcrops, the so-called "aureoles" that display high gamma radiation. Those aureoles are distributed over a wide area in regions with dominant chemical weathering. Reconnaissance uses the propensity of the elements described above to be transported by subterranean waters to relatively large distances from deposits and outcrops. Fairly extensive mechanical aureoles (alluvial and eluvial deposits, mainly pitchblende and uraninite) are produced in areas of mechanical weathering and primary uranium quartz ores (Melkov, 1956).

In depth prospecting, an estimation work takes place in districts with radioactive anomalies and its surrounding areas as well. The geological, geomechanical, hydrogeological, geomorphological, and other peculiarities of the area that contains the anomalies are of great importance. Proper prospecting is important in the surrounding areas because the anomalies change in relation to the original ores, and also because not all ore-bearing districts or deposits are distinguished by high surface radioactivity, which is characteristic of areas of concealed ores and districts buried by sediments. It is important to know the general geological laws of uranium distribution, the characteristics of common types of uranium deposits, and the behavior of uranium in a weathered crust for efficient and accurate prospecting and evaluation of radioactive anomalies in neighboring districts (Melkov, 1956). The following points are brief notes presented by Melkov (1956) to be understood and taken into consideration in the case of magmatic uranium deposits (The infrequent deposits containing davidite (iron titanate containing 7% uranium) are not considered here):

1. Magmatic deposits of uranium are formed primarily in competition with the latest representatives of the magmatic division of granitoid

magma, highly enriched with silicon and volatile elements, less with alkalis and very low in magnesium, iron, and calcium.

2. Usually deposits of uranium minerals are not found in massive, broad variety and regional abruptions, but rather in associated fractures. These fractures, results from branching, blending, and collisions with dykes of volcanic rocks. Quartz, and other veins or unmineralized fractures in, as well as to their junction with other host rocks that are comparatively rich in iron oxides. The mining of uranium ores is also likely to occur in areas close to the branching of extensive fractures. Uranium ores are formed in fractures distinguished by their many replacements.
3. Uranium has an ability to form the largest ore bodies in primarily "hard rock" fractures such as various metamorphic rocks (such as schists, quartzites, hornstones, and gneisses), granites, granodiorites, and acid effusive rocks. Metamorphic and extensively metamorphosed rocks are the most attractive, particularly those rich in diverse divalent iron minerals such as magnetite iron-bearing varieties, amphiboles, and chlorites. The largest bodies of uranium ores are found in intrusive masses within effusive and volcanic strata, typically in hydrothermally altered zones containing talc, clays, chlorites, and pyrophillites and even in parts of the dykes of simple volcanic rocks (diabasic porphyrites). Magmatic reserves of uranium are lacking in strata of normal sedimentary rocks except in the parts adjoining strong rocks.
4. Mineralization of reserves of uranium is not an instantaneous process but rather goes through some pulsational stages. Most deposits begin with a pneumatolytic period and end with a hydrothermal phase, going through six or more stages of destruction and replacement. Some deposits have initial formation phases characterized by metasomatosis and formation of pores and caves, an example of which are albites. The formation of uranium ores is generally related to one of the final phases and it is independent in form and often in space. This condition is the most positive one to the development of the highest uranium deposits, other conditions are of less but equal significance. The comparatively small number of chemical elements associated with uranium also distinguishes this stage of uranium. The important ones are silicon, iron, and oxygen, and always found with the uranium. Carbon, calcium, and fluorine are found in large quantities but not all the times. Aluminum, magnesium, sulfur, and the rare earth elements are the elements that occur less commonly and in smaller quantities. At last, zinc, arsenic, lead, and copper occur the least frequently and in very small amounts. The

above-mentioned elements form various associations under different conditions such as: pitchblende and silica (metacolloid); pitchblende-quartz-hematite; pitchblende-quartz-calcite (or dolomite); pitchblende chlorite (thuringlte and daphnite), often with quartz; pitchblende-fluorite-calcite and a few others. Besides these main minerals, other minerals also occur in small quantities such as pyrite, sphalerite, galenite, arsenopyrite, and enargite. The rare earth elements and often lead, silicon, aluminum, and a few more elements do not exist naturally but instead are found in metacolloidal pitchblende structures.

In the event of spatial changes or the overlapping of the uranium mineral formation cycle on minerals produced in one or more preceding ages, regions with vein deposits tend to be characterized by the complex multi-mineral composition of uranium ores, followed by significant, specific alterations in areas around the veins. Examples of multiple ores may also be cited, excluding the above-mentioned uranium minerals and the mineral composition of host rocks such as uranium ores with large quantities of iron, zinc, and lead sulfides; sulfides and sulfur salts of copper, barite and bismuth; copper, iron, and molybdenum sulfides; wolframite, iron, copper, lead, and zinc sulfides, bismuth, sulfur salts and barite; alkali (sodium) silicates and alumosilicates, magnetite, native silver, hematite, arsenic, bismuth, arsenides and diarsenides of cobalt and nickel, and iron and copper sulfides; iron sulfides, hydroalumosilicates (hydromicas and montmorillonite). The variability of uranium deposit mineral composition seems to be related to its size and the richness of the ore body. It has been noted that sulfides are typically absent from large uranium ores. The presence of sulfides is limited to the smaller deposits.

5. Host rocks are not altered during the uranium stage of mineralization. However, the formation of hematites (red coloration) and chlorites (green coloration) is still possible. The other most closely associated formations are carbonates, quartzes (hornstone-like quartz), argillites (usually a mixture of varying quantities of hydromicas, montmorillonite, galluasite, and kaolinite), and fluorites.
6. The behavior of the uranium and its daughter radioactive elements (ionium, radium, and radon) varies significantly in the development of oxidation zones, depending on the mineral composition of the primary ores and conforming to the chemical characteristics described above. Due to the predisposition of hexavalent uranium compounds (which in all cases are formed in the pitchblende and uranium "blacks" oxidation

zone) to dissolve, the above distinctions slowly disappear in old oxidation zones with steady and prolonged erosion and thus leave the deposit. According to Melkov (1956), sulfides enhance the uranium mobility in oxidation zones and the following distinctions are usually noted in young oxidation areas with lower baseline erosions:

(a) In sulfide deposits, all of the uranium in the pitchblende dissolves completely which leads to the development of secondary vertical zonality in the distribution of hypergenic uranium minerals.

(b) In the cementation region, pigments of revived uranium are contained in the active oxidation sub-zone (recent oxidation sub-zone of the cementation zone). Residual uranium pigments and U6 + solutions fall out due to the oxidation and dissolution of the pitchblende. Higher in the vertical profile of the oxidation zone, there is a sub-zone void of uranium minerals followed by a sub-zone of uranium micas of the phosphor-calcium type. In this sub-zone, the arsenical and copper uranium micas presence is positively associated with the occurrence of copper sulfides and arsenides in the major uranium ores. In this instance, the arsenical and copper micas (zeunerite, torbernite and uranospinite) are found in the bottom part of the sub-zone and the phosphor-calcium otenite comes at the top. Finally, uranium silicates (uranophane) come at the uppermost part along with minerals that include an admixture of uranium (uranium-bearing hyalite and calcite). Uranium-bearing limonite is dispersed all over the oxidation zone.

(c) A variety of uranium silicates and hydroxides, which are pseudomorphs of tiny or microcrystalline structure (gummites), occur in the pitchblende oxidation zone that contains little or no sulfides. The different "blacks" uranium and micas of uranium are poorly formed. Uranium micas increase at the top of the oxidation zone that shows that vertical zonality also governs the division of secondary uranium minerals; however with a lower level of clarity and differentiation.

According to Melkov (1956), other rules must be followed in exploring sedimentary deposits and surrounding districts, and assessing their anomalies as well:

1. Uranium deposits usually occur in concentrated narrow strips (few hundreds of meters) in a much larger area of low uranium content. If the strip with the highest uranium is located, it would usually be tracked along a shoreline at a close distance (from a few tens of meters to a few

kilometers). Coal basins could also be rich in uranium deposits; however the presence of uranium bodies in coal and in a number of other caustobiolites (such as asphaltites) is apparently due to epigenetic rather than syngenetic processes.

2. In normal sedimentary rocks, uranium is typically found in mixed oxides in a sprinkled state and frequently in concurrence with organic matter (mainly humic) contained in phosphates (such as concretions of phosphorites, fossilized skeletons, and integumentary parts of fish of phosphate composition) and carbonates (such as shells and calcite oolites). Uranium is rarely associated with independent minerals, such as pitchblende and "blacks," that are usually found in tiny (often microscopic) deposits. Pyrite could form up to 30% of uranium deposits that also contain relatively large quantities of sphalerite, galenite, sulfides of vanadium, barite, and molybdenite. Phosphatic rock components frequently contain fluorine, rare earth elements, scandium, and a few other elements. Unoxidized sedimentary rocks with high uranium content are not distinguished by limonite and reddish ores but rather in gray to gray-green ones.
3. Uranium is no longer associated with organic matter in highly metamorphosed sedimentary rocks. In these rocks, the organic matter changes to solid and well-carbonized compounds (such as anthraxolites), and uranium transforms to pitchblende and other minerals distinguished by epigenetic distribution in the seam of the ore.
4. The oxidation zone of sedimentary deposits with low uranium content lacks secondary uranium minerals except some uranium-bearing limonites, and seldom, separate laminae of otenite and torbernite. On the other hand, in the oxidation zone of sedimentary deposits with high uranium content, substantial amounts of phosphates (alkali and alkali-earth micas) and sulfate-carbonates (Schroekingerites) are generally found.

When vanadium is abundant, secondary vanadates (tyuyamunite and carnotite) are found in uraniferous sedimentary strata. In most of the cases, secondary uranium minerals come with iron sulfates and phosphates, limonites, gypsum, magnesium, and aluminum.

CHAPTER 3.4

Methods of surveying

Various methods of surveying were developed in the last few decades and include air-based regional radioactivity survey, car-borne radioactivity survey, radio hydrogeological survey, geochemical survey, and geophysical surveys. These methods, in addition to other methods used directly in the field such as emanation survey, uranometrical survey, luminescent survey, and geobotanical survey are discussed in the subsequent sections.

Regional geological survey

Melkov (1956) believes that the prospecting for uranium should be built primarily on good knowledge of the geological and mineralogical nature of the areas of interest. This, in conjunction with good knowledge of the factors controlling the distribution of uranium in various environments, will identify the key indicators of the existence of radioactive ores on the basis of existing promising geological features and their characteristics. The main geological features or environments of interest at this stage include the following:

- Regions with major acidic igneous rocks.
- Regions with vertical major dykes or sills.
- Regions with major unconformity belts.
- Thick clastic sand deposits containing magmatic rocks, or rich in clay or organic matter.
- Zones of oxidation in continental sediments deposited in the marine, river, or delta environment.
- Regions of phosphatic rocks or black shales.

Air-based regional radioactivity survey

Regional radioactivity surveys along with regional geological surveys are used in most countries to find radioactive materials and together represent the first phase of surveying. The main regional radiometric survey is the regional airborne radioactivity survey. This survey is very rapid, can cover large areas, and very effective in detecting radioactive anomalies. It uses aircrafts fitted with highly sensitive digital radiometric monitoring devices at a short time interval (Geiger and scintillation counters), which register not only a field's total gamma activity but also its spectral composition. The surveying planes and helicopters are also equipped with tools for altitude measurements and aerial photography as well (Melkov, 1956). The units of gamma radiations emitted from rock formations containing radioactive materials lead to determining the radioactive anomalous zones. Thereafter, these zones are subjected to detailed surveys as will be described in the subsequent sections.

Melkov (1956) found that the most successful airborne radioactivity surveying is based on the results of the geological studies of the target area, with allowance for the geological peculiarities of the uranium ore deposits, and carried out by geologists. The scale of the airborne survey is chosen based on the anticipated nature of the uranium deposit, outcroppings, topography, and understanding of the geological structure of the target area. The survey is carried out in equidistant parallel courses that depend on its scale and the height of flight, and generally pass across the target uranium deposits and levels (The path of the air courses in very disconnected regions must be chosen according to the geographical profile of the region.) (Fig. 3.4.1). A course is generally 30–40 km in length and the number of courses is inversely related to the height of the flight. The height

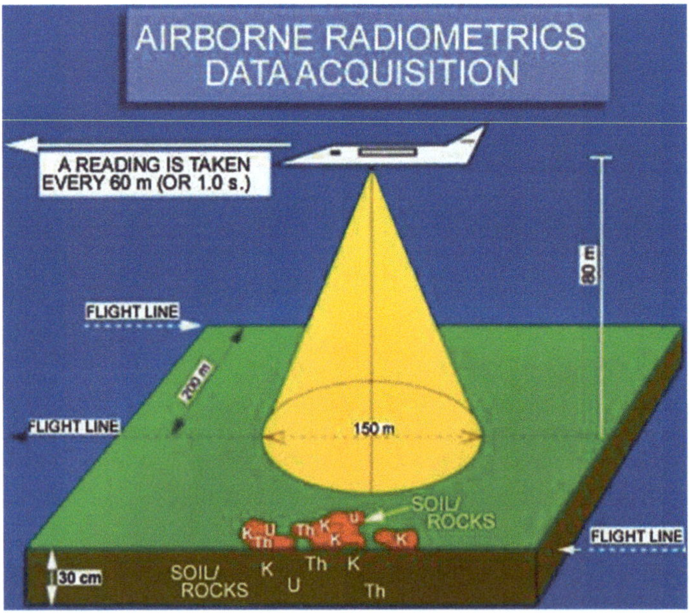

Fig. 3.4.1 ***The technique used in the airborne radiometrics data acquisition (Department of Primary Industries, Australia, 2008).***

of the flights ranges from 15 up to 120 m and depends on the radiometric and aerial techniques used and the geography of the surveyed area as well. The outcomes of the air-based radiometric survey along with the results of the geological survey are used to produce a radiometric aerial map displaying the gamma activity isolines. The detected radioactive anomalies are represented by absolute gamma radiation released from deposits and outcrops of uranium, concentrations of thorium, and destructive or effusive oxidized stones. Districts by notable gamma activity are characterized by an aerial gamma spectrometry that links the size, form, intensity, and placement of the gamma anomalies to the geographic description and fundamental aspects of the geological structure. The sizes and shapes of gamma anomalies of igneous rocks and effusive strata depend on their respective areas. These gamma anomalies emit almost the same level of gamma radiation over the whole anomalous area. The gamma anomalies of deposits and outcroppings of uranium and thorium form a relatively sharp gamma peak in local anomalies and are often noticeable against the background of an anomalous district (Melkov, 1956).

In addition to the airborne radioactivity survey, there are three different types of regional radiometric surveys: handheld radiological survey, total gamma-radiation survey, and airborne gamma ray spectrometric survey.

- *Handheld radiological survey:* This survey uses the intensity of radioactivity of rocks to locate distinct radioactive anomalies. It is also used to spot targets for future detailed surveys to determine the sources of radioactive anomalies that could be uranium, thorium, or potassium deposits.
- *Total gamma-radiation survey:* This survey records the intensity level of gamma ray energy. Different radioactive elements emit gamma rays with various energy levels, and this helps in locating the areas whose radioactivity comes from uranium ore deposits.
- *Airborne gamma ray spectrometric surveying:* Airborne gamma spectrometry is the main approved uranium prospecting method with global applications for geological mapping, mineral mining, and environmental surveillance. Due to its high cost, this method is often used for specific areas after conducting the total gamma radiation survey. This survey selects smaller areas with the highest radioactivity level for a direct ground survey. However, this survey may also cover large areas of low radioactivity associated with the presence of raw uranium deposits.

For instance, when potassium K^{40} decays, it produces argon Ar^{40} and emits gamma-ray with an intensity of 1460 KeV electrons. Most of the new aerial radiometric survey devices can detect natural gamma ray spectra between 0 and 3000 KeV and have 255 channels. As a result of the three methods mentioned above, a regional map is created to show the intensity of radioactivity of different geological formations, in addition to an isorad map that shows areas of equal intensity of radiation.

Car-borne radioactivity survey

The car-borne radioactivity survey supplements the aerial method. Once the districts of high gamma activity are detected through the air-based survey, and if the terrain permits, a detailed study could be conducted using this method. This survey could also be used to cover the gaps of the aerial survey. Also, the car-borne radioactivity method is used as a large-scale survey in regions of relatively small areas. However, this method is significantly limited by the availability of roads in the target area. Advanced car models are equipped with technologically advanced radiometric devices such as the ones used for the airborne radiometric survey (Melkov, 1956).

Radio hydrogeological survey

Uranium ore deposits could be buried in areas filled with alluvial deposits that will absorb gamma radiation and consequently could be discovered by a radio hydrogeological survey. This survey, as described by Melkov (1956), is used mainly for assessing "the accumulation of radioactive elements, minerals, red-ox potential, radioactive and non-radioactive gaseous materials resulting in the decay and disintegration of uranium atoms, and a variety of other amounts contained in water samples obtained from both natural and artificial water sources including flowing water." The collected water samples are analyzed in the field using equipment that enable such swift analyses. Subsequently, radio hydrochemical maps are developed on a geological basis with the appropriate scale after analyzing the radio hydrochemical and hydrogeological studies of the target area (Melkov, 1956).

Geochemical survey

The exploration objectives, expected target area size, and logistical arrangements must be taken into consideration in the design and sampling strategy of a geochemical survey. All available geological and geophysical data should be incorporated in the design of a surface geochemical survey to obtain the best potential outcomes. This section reviews the available methods for conducting surface geochemical surveys and discusses how to design a survey of maximum effectiveness.

A geochemical surveys main objective is to obtain accurate chemical knowledge from a given area rocks or soils. Samples of rocks, sediments, or soils are collected and analyzed in chemical or geochemical laboratories to detect certain chemical elements or compounds. Geochemical analyses are used in mineral exploration to locate the areas anomalous in the commodity sought, or in elements associated with it (www.wikepedia.com). There are two types of geochemical surveying techniques: direct or indirect. Direct techniques study the little hydrocarbons contained in the porous soil, especially those accumulated on the fine-grained part of soil or enclosed in the soil cements. Indirect geochemical techniques identify the alteration to soil, sediment, or vegetation caused by the seepage of hydrocarbons (Shumacher, 1999). Fluvial sediments were usually targeted in regional geochemical exploration to locate mineralized water. Regional surveys may use a small number of samples collected from the target area, sometimes in the order of one sample/100 km^2, due to their large scale. The next step in the geochemical survey is the systematic distributed sampling from soil in areas where soil is amenable to soil geochemistry. Shallow drilling is another sampling method in areas covered by alluvium or heavily distorted by human activities.

The data obtained from the geochemical analyses are examined for anomalies (single or multiple elements) associated with mineralization. The geochemical anomalies are usually checked in the field and normalized against the regolith type and landform to avoid misleading information caused by weathering, erosion, and landforms. Geochemical anomalies could be misleading and associated with low-grade mineralization. To avoid that, geochemical anomalies are drilled to assess the economic concentrations of mineral deposits. Chemical analyses of rocks and plants may help in locating buried ore deposits. Some chemical elements are associated with specific minerals. For example, arsenic and antimony are associated with gold deposits so their presence may indicate their coexistence with gold. Also, tree buds could contain some chemical elements that indicate the presence of specific mineral deposits (www.wikipedia.com).

Geophysical surveys

Earth materials react differently when subjected to gravitational, magnetic, electrical and electromagnetic fields, or seismic waves in a way that reveals their physical characteristics. Geophysical surveys may be conducted from the ground surface, air, sea, or space at various scales. The data obtained from these surveys are usually processed by computer software packages to produce a large variety of colored images that display different geophysical properties of the rocks. The geophysical surveys discussed in this section are gamma-ray, electromagnetic, induced polarization (IP), and remote-sensing geophysical surveys.

Gamma-ray method

Airborne gamma-ray spectrometry is currently the most reliable tool in uranium exploration. It is also used in geological mapping, mineral exploration, and environmental monitoring. Gamma-ray methods (Durrance, 1986) use scintillometry to detect the three natural radioactive elements: potassium, uranium, and thorium. Multi-channel spectrometers measure the abundance of these radioactive elements. Gamma-ray surveys are extensively used in uranium exploration because they directly identify the presence of the natural radioactive elements. Among the three natural radioactive elements, thorium is usually the most static one and has geochemical characteristics analogous to those of the element zirconium. Thorium and uranium are more abundant in felsic rocks and their content increases with alkalinity.

Dickson (2004) noticed that in the last few years, the aerial gamma-ray surveying is using more advanced techniques and that is reflected in the quality of data collected. The main techniques currently used include advanced aircraft equipped with a global positioning system which allows very low-altitude surveying (10 m) with course spacing as narrow as 10 m, advance in noise-reduced surveying helps in collecting more data, and highly developed techniques to reduce the haziness caused by the uncollimated detectors and deconvolution methods.

According to Asfahani et al. (2006), airborne gamma-ray surveying was successful in the exploration of uranium associated with phosphate deposits in Syria. They used airborne gamma-ray spectrometric data with a factor analysis technique and found that this method is very effective in representing the obtained geophysical data and evaluating the fundamental factors that direct their variances. As mentioned above, gamma-ray spectrometry quantifies the natural radioactive elements and consequently, may provide geo-environmental data pertaining to radiation dose and radon potential.

The occurrence of uranium and/or potassium may indicate a possible release of dangerous materials from ore or waste piles. It is also important to mention here that uranium mobilization speeds up by the oxidation of sulfide minerals (Hoover et al., 1995).

Electromagnetic method

Griffiths and King (1981) described the electromagnetic surveying as follows. The electromagnetic prospecting techniques are centered on the analysis of the magnetic fields correlated with alternating currents caused by primary magnetic fields in the subsurface conductors. Generally, an AC current is transmitted through a transmitting wire, with a frequency between a few hundred and a few thousand hertz. Therefore, eddy currents are produced in each and every conductive ore body. The magnetic field at the surface is the product of the transmitter-related secondary field formed by this induced current and the primary field. There will normally be a difference between the primary and second fields, in both direction and phase. The current induced in the horizontal receiving coil is dependent on the magnitude and frequency of the vertical portion of the resulting field. Therefore, magnitude and phase can be determined when compared with the primary field.

The airborne electromagnetic (AEM) systems were first used in Canada in the early 1950s for base metals exploration. The demand for AEM surveys increased substantially until the late 1970s with a coverage of 600,000-line km/year. During the 1960s and 1970s, 18 systems were used by Canadian companies that conducted approximately 80% of AEM surveys in the Western world (Palacky, 1986). AEM surveys continued to be the most used airborne geophysical surveys, and consequently the main source of revenues even when minerals production decreased. Currently, remote-sensing methods are not capable of substituting AEM surveys in mineral prospecting and many other applications because surveying profound geological structures requires the low frequencies (between 100 Hz and 50 KHz) of the AEM method. Also, satellite sensors are unable to record the signal as accurately as the ones of a low flying aircraft (Palacky, 1993).

According to Palacky (1993) the electrical conductivity of the ground greatly affects the AEM records (Fig. 3.4.2). Other properties such as magnetic susceptibility could also influence the AEM surveys especially if the target area contains magnetic minerals.

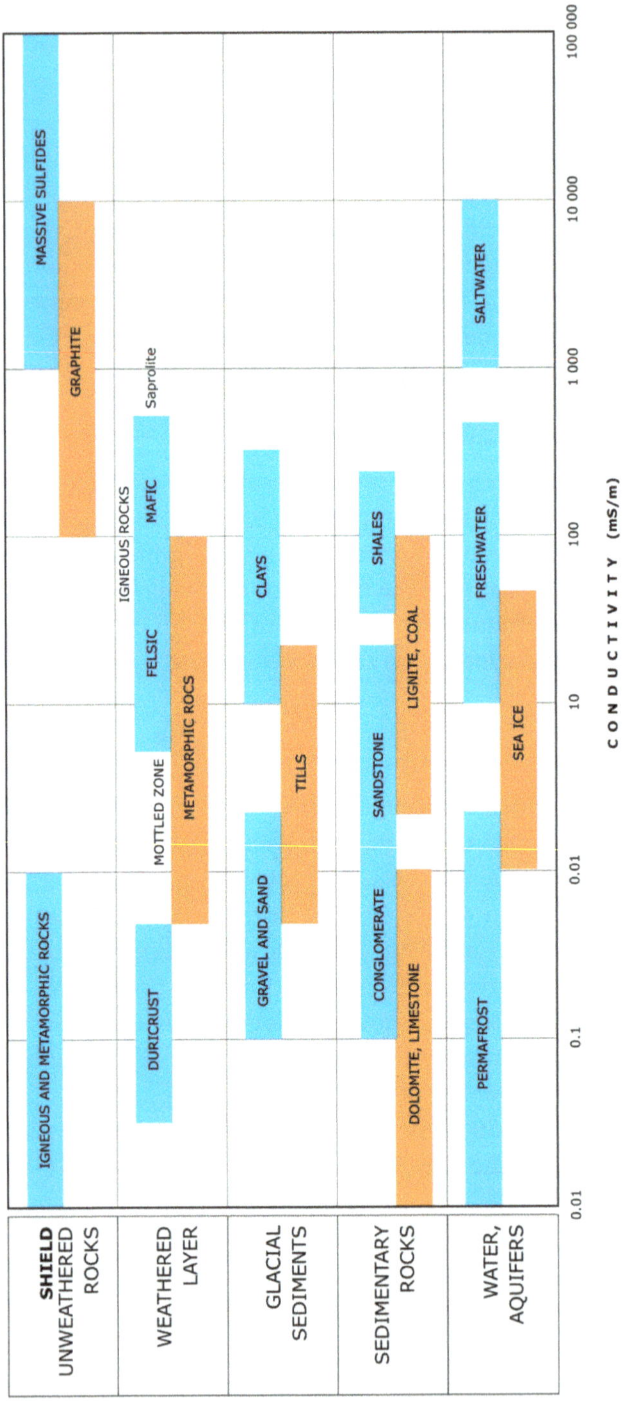

Fig. 3.4.2 *Typical ranges of conductivities of common earth materials.*

Massive sulfides of volcanic origin, where most AEM surveys are conducted, are known by their high conductivity (usually more than 1 S/m). However, sulfides of low or no conductivity have been mentioned in the literature (Palacky, 1988). It has been noted that the conductivity range of the mineral graphite covers a considerable part of its sulfides counterpart. In shield areas, massive sulfides are found in very resistive igneous and metamorphic rocks (Palacky, 1993).

AEM surveys can cover vast districts and determine the local areas where ground prospecting would take place. They are very useful in spotting low-resistivity and high-conductivity massive sulfides, the usual target in mineral exploration. Electromagnetic methods can also map lithological and structural features which might be useful in mineral exploration and geo-environmental studies as well (Palacky, 1986; Hoover et al., 1991). In addition to mineral exploration, electromagnetic methods are progressively more used in groundwater exploration.

Induced polarization (IP) method

The IP method quantifies polarizable minerals (metallic-luster sulfides, clays, and zeolites) found in the water that occupies porous rocks. This method can spot polarizable minerals only when the pore water covers their active surface, and consequently dispersed sulfide minerals are detected more often than massive sulfide deposits whose IP could also be measured. IP is widely used in detecting scattered sulfide ores with little volumes that may be as small as 0.5% of the total volume of sulfide minerals with metallic luster (Sumner, 1976). In addition to detecting and measuring polarizable minerals, this method has been used in geo-environmental studies, especially in delineating clay aquitards that can control the runoff of the mine's wastewater (Hoover et al., 1995).

Remote-sensing method

"Remote sensing involves techniques that use images collected near the electromagnetic spectrum in the ultra-violet, visible, and infrared bands" (Hoover et al., 1995). Thermal infrared observations form part of remote sensing as well. Remote-sensing data are usually displayed in a digital image format to be easily processed. Iron hydroxide minerals, silica, and clay alteration have relatively broad spectral responses. Remote sensing is also used in geo-environmental studies to outline surface distortion patterns (Knepper, 1989) and to spot anomalous metal presence in soils characterized by unusual vegetation patterns (Birnie and Francica, 1981).

Other surveying methods

Once aero-radiometrical and radio-hydrochemical anomalies are spotted in a target area, detailed geological, petrological, stratigraphical, and mineralogical studies will be conducted on its surface rocks using methods that deal with the characteristics of uranium and other radioactive elements. Melkov (1956) described these methods as simple and mentioned four of them that are discussed in the subsequent sections: the emanation survey, uranometrical survey, luminescent mineral survey, and geobotanical survey. All these methods are analytical, conducted directly in the field, and based on common exploration rules. To carry out these surveys, a grid of parallel courses is set up mainly along transverse lines with a scale based on the geological and geomorphological characteristics of the target area and its anticipated type of deposits. Samples of rocks, soil, ground gases, and vegetation are collected along these lines at a constant interval of few meters. Luminescence is continuously surveyed, and emitted beta and gamma rays are recorded. The outcomes of these surveys are presented in different forms such as profiles, maps, and plans.

Emanation survey

The emanation survey could be carried out in a few minutes and is usually conducted in areas buried under a thick cover of sediments, and hence capable of exploring areas at a relatively substantial depth. The main objective of conducting an emanation survey is to measure the concentration of the gaseous radioactive elements' radon and thoron, produced by the radioactive decay of Ra and Th that are the daughter elements of uranium and thorium. The concentration of radon and thoron is measured in air samples collected at a depth of 0.8–1.0 m from the ground surface using particular sampling drilling tools. Air samples that need to be taken from a greater depth are collected from special drilled wells (Melkov, 1956).

Uranometrical survey

We have mentioned before that uranium could be contained in alluvial sediments near aureoles. This survey measures the uranium content in this kind of deposits. Samples are collected from drilled wells, pits, and trenches at various depths. When dissolved in sodium fluoride, uranium fluoresces under ultraviolet irradiation and that enables the measurement of the uranium content through a luminescence analysis. The intensity of uranium fluorescence at low concentrations ($1–5\times10^{-2}\%$) is proportional to its concentration (Melkov, 1956).

Luminescent mineral survey

This survey is always conducted at night and along the whole courses by continuously irradiating the surface of outcrops with ultraviolet light. Few types of secondary uranium and uraniferous minerals such as uranium micas and uranium-bearing hyalite and calcite fluoresce in ultraviolet light and thus, points of luminescence are spotted to be further studied afterward during the day (Melkov, 1956).

Geo-botanical survey

The geo-botanical survey could be conducted in two ways that are both capable of locating buried uranium deposits at a maximum depth of 25 m: (1) burning some plants and screening their ash for radioactive (mainly uranium) and other elements that are directly associated with radioactive ore deposits, and (2) searching for particular plants whose growth is influenced and chemically altered by the presence of certain chemical elements exclusively contained in uranium ores (Melkov, 1956).

Using some or all of the methods of surveying mentioned above may help in locating the radioactive anomalies and determining their nature. Melkov (1956) believes that after finding partially preserved uranium mineralization, the size and scope of the mineralization should be determined using suitable methods. It is also crucial to consider the special characteristics of the structure of uranium ore deposits and their oxidation zones, especially those caused by the geographical and hypsometric location of these deposits and their surface profiles. Finally, to establish a well-organized exploration process, it is important to consider the likelihood of a substantial change of the radioactive balance, which is usually governed by the different sub-zones rules of the hypergenetic zone.

Exploration drilling

This step comes after locating uranium high-potential areas. Exploration drilling is the mean by which geologists take rock samples from few meters below the earth surface to depths exceeding 1000 m to study the geological and geochemical characteristics of the sub-surface geology. A mineral exploration system typically uses drilling rigs with a wide range of drilling techniques, depending on the conditions of the region and other factors. (New South Wales Minerals Council Ltd, 2008).

Bulk sampling

This step may be undertaken to achieve a sufficient volume of materials to ensure relevant and useful test work to be taken on, often as part of the analysis of pre-feasibility and feasibility for mining. Bulk sampling actually requires trench excavation or test pits from which a large sample volume is extracted. Uranium deposit is identified using exploration methods to estimate the uranium amount that could be extracted from it at an economical cost. Proven uranium reserves are the amounts predicted to be recoverable at present economic costs (New South Wales Minerals Council Ltd, 2008).

CHAPTER 3.5

Exploration case studies from the Middle East

Some of the surveying methods discussed in previous sections of this chapter were applied in two target areas for uranium exploration in the Eastern Desert, Egypt, as a part of the modern methodology used for uranium exploration. The studies conducted on two areas of Egypt—El-Missikat area in the central Eastern Desert and the Kab Amiri area in the Eastern Desert—are discussed in the subsequent sections.

El-Missikat case study

Rabie et al. (1996) used ground gamma-ray spectrometric survey, the Gad-6 (Scientrex Ltd, Canada) and the gamma ray sensor (model Gsp-4), in their uranium exploration in El-Missikat prospect area in the central Eastern Desert of Egypt. Thirty-four parallel profiles perpendicular to the strike of the mineralized veins were chosen in the target area. The authors set a base line that has the same direction of the veins. The distance between two consecutive profiles was set to 50 m, while the stations put along the profiles were separated by a distance of 20 m. Therefore, the average lengths of profiles (coded 650 E to 1000 W) were in the order of 1 km. Elkattan et al. (1996) evaluated several common techniques often employed during uranium exploration activities in the El-Missikat area, central Eastern Desert, Egypt. They used self-potential (SP), IP, and horizontal loop electromagnetic (HLEM) surveys.

The method of self-potentiality (SP) is based on the natural potential difference between any two locations in the ground (Sato and Moony, 1960). These potentials are mostly constant but occasionally fluctuate irregularly and are related with subsurface electric currents. Electrochemical reactions in underground rocks and/or in materials found in these rocks produce these potentials. There are two types of these potentials: First, the minor positive or negative potentials with an intensity of a millivolt to a few millivolts, second, the major and generally negative mineralization potentials that may have intensities of hundreds of millivolts (Telford et al., 1990). The large mineralization potentials occur over sulfide and graphite zones. The SP method is economical, easy to conduct in the field, and extensively used in mineral exploration (El Kattan et al., 1996).

El Kattan et al. (1996) described the settings of their SP survey of the El-Missikat area as follows. "SPs are difficult to detect across surface rocks with high resistivity. Hence a high impedance voltmeter instrument would be used. Accordingly, the meter used in the present study, along with two non-polarizing electrodes, is an Auto-Ranging Digital Multimeter model ME-540. These non-polarizing electrodes are typically made of porous pots which contain a saturated copper sulfate solution. The fixed electrode technique was chosen for conducting the SP survey in this field of study. In this methodology, at the selected base station one electrode is set, and the other moves along the profile. All the individual base stations (base for each profile) were reduced relative to the zero value of the selected general base station. The sample interval of observation along the profiles was 20 m and the distance between the profiles was 50 m. After reducing the SP data, the value of each profile's base station was subtracted or added to the value of each station along the same profile." A contoured map was drawn out of the resulting reduced values (Figs. 3.5.1-3.5.3).

Finally, in this research, Elkattan et al. (1996) carried out an electromagnetic survey with the same system used in their whole work which mainly consists of a transmitter and receiver. Their models are TM-2 and IGS-2/EM-4, respectively, and they were both provided by the Canadian company "Scintrex" (Fig. 3.5.4). This system is capable of carrying out an HLEM survey. The system works by transmitting primary currents with various frequencies and receiving them as secondary signals at every frequency, and hence provides data about conductors at different depths. Deep conductors could be explored with lower frequencies.

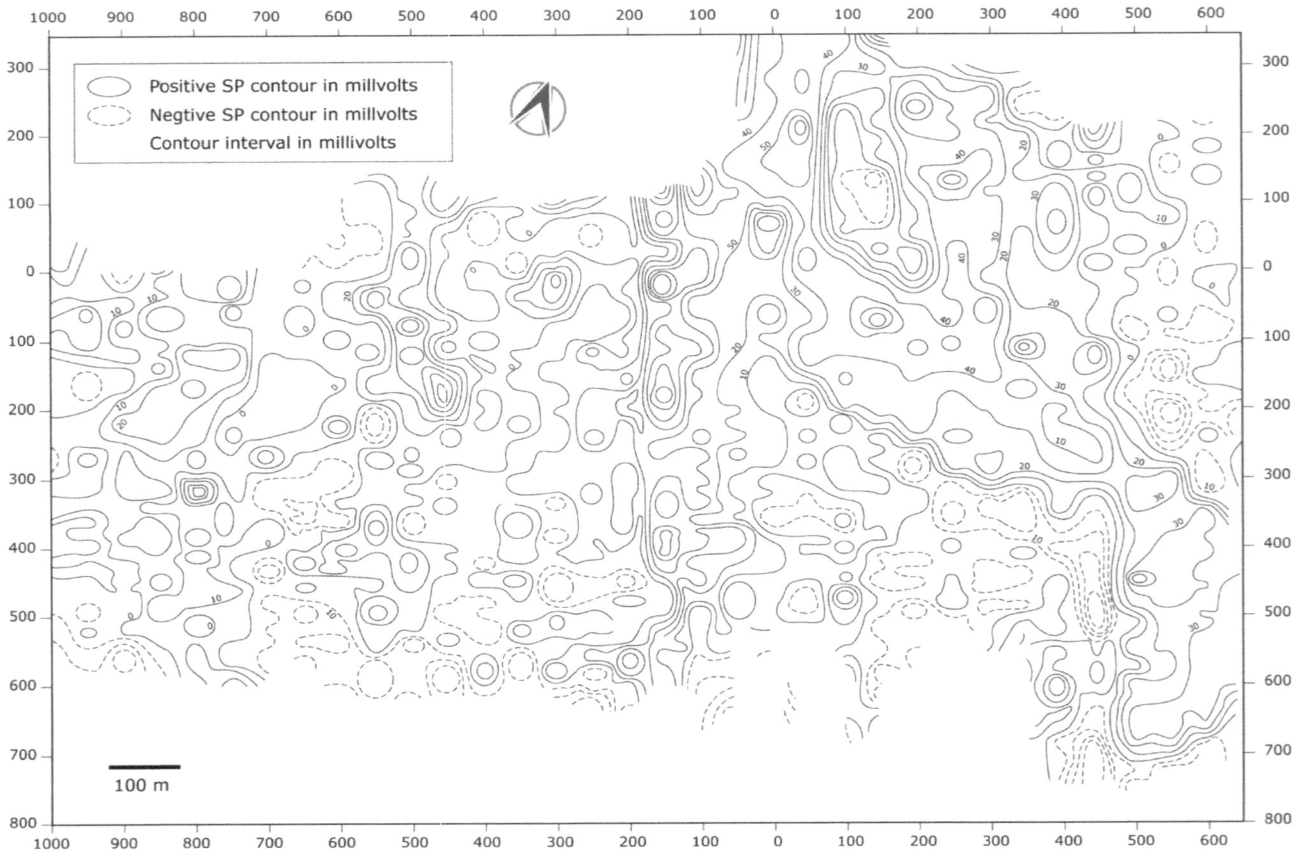

Fig. 3.5.1 ***Ground SP map of El-Missikat prospect, central Eastern Desert, Egypt (El Kattan et al.,*** 1996).

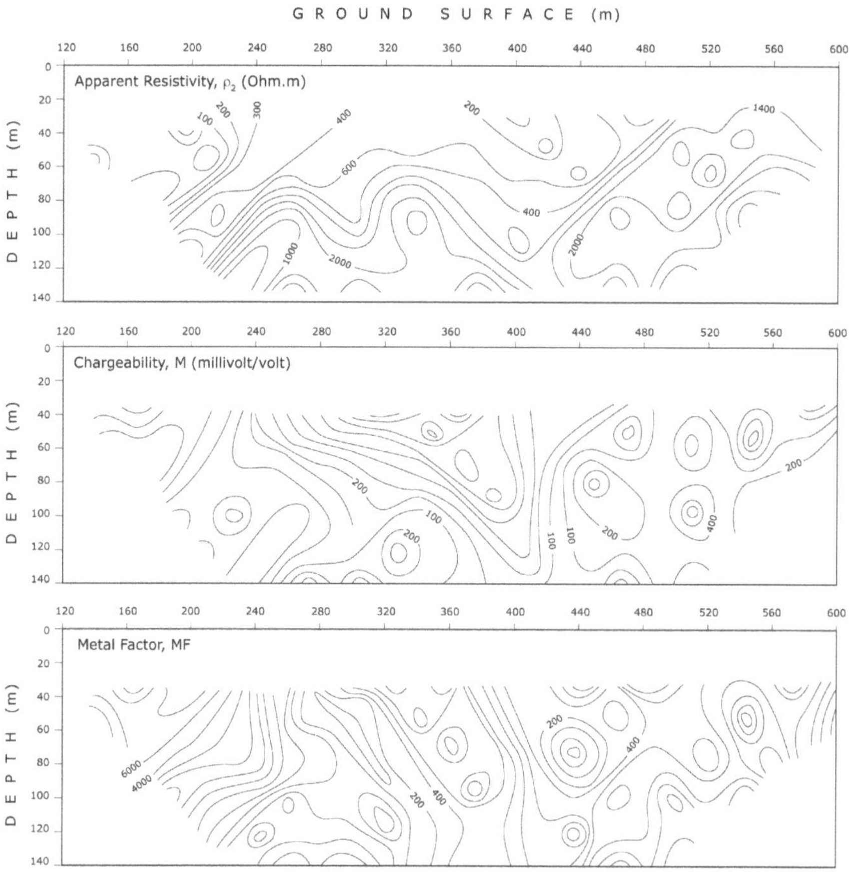

Fig. 3.5.2 ***IP 450E profile pseudo-sections, El- Missikat prospect, central Eastern Desert, Egypt (El Kattan et al.,*** 1996).

The transmitter and receiver must be separated by a fixed distance when conducting an electromagnetic survey. The distance in their survey was set to 140 m. The survey was carried out along three special profiles in the target area to collect data about the conducting zones, particularly those located near shear zones and have geological structures such as dykes or veins. The selected frequencies were 112.5, 337.5, 1012, and 3037 Hz. At every frequency, the transmitter emits a primary signal that has an in-phase and out-of-phase modules.

These primary signals are collected by the receiver that computes the percentage ratio of their modules. The angle formed by the intersection of the plane that contains the transmitter and receiver and a horizontal plane, called the tilt angle, must be considered in these calculations.

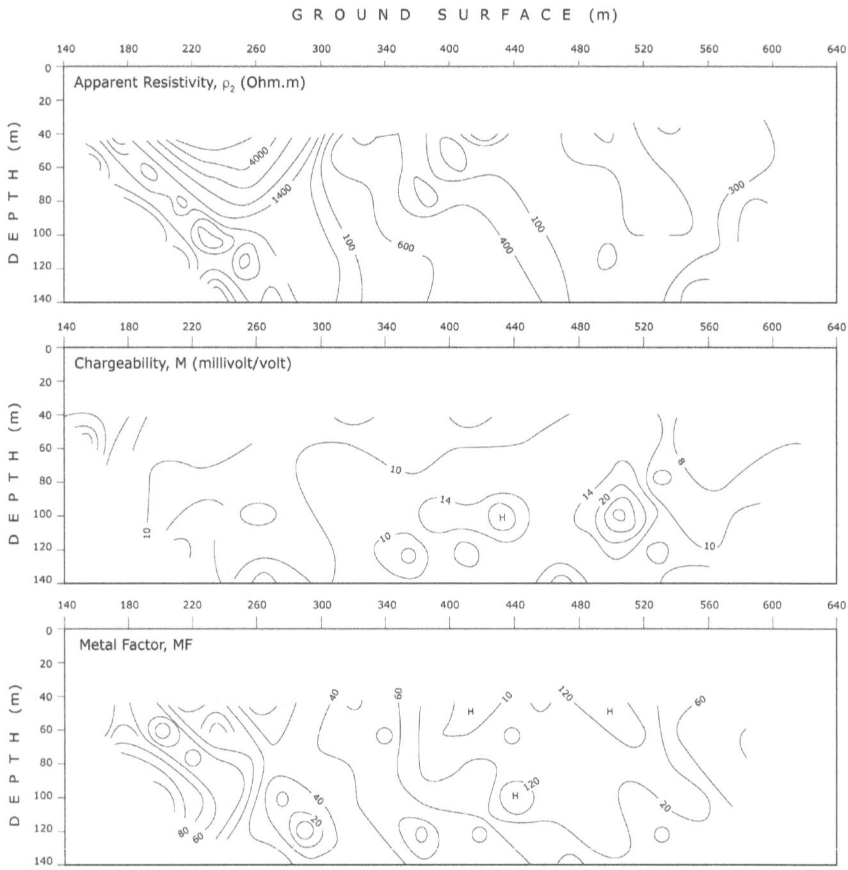

Fig. 3.5.3 ***IP 500E profile pseudo-sections, El- Missikat prospect, central Eastern Desert, Egypt (El Kattan et al.***, 1996).

Kab Amiri case study

A study was financed and technically supported by the International Atomic Energy Agency and the Nuclear Materials Authority to locate the Egyptian granitoid type in the Eastern Desert of Egypt, especially those that have similar characteristics to the granitoid rocks in Europe which contain economical uranium deposits. First, the data gathered during previous studies were reevaluated, taking into consideration the present geological concepts of plate tectonics and igneous petrology. The next step was to conduct a reconnaissance study to determine on a case-by-case basis the essential studies and the necessary field and geochemical laboratory analytical tests. As a result of this reconnaissance study, the granitoid types with the

Fig. 3.5.4 ***The hardware onboard instruments provided by the Canadian company "Scintrex".***

highest uranium potentiality were identified and subjected to more detailed geological, geophysical, and geochemical studies and alteration detailed mapping. These studies revealed two areas with high uranium potentiality: the Kab Amiri granitoids (central eastern desert) and the El-Sela granite (south eastern desert).

During the third step, geological, spectrometric, and mineralogical in-depth studies were conducted in the first area (Kab Amiri) on a 500 m grid. More specific and technical investigations were undertaken in the last stage based on the refined conclusions of the previous three steps. These final studies consisted of the following: Demarcating peripheral phase of granite with high uranium and thorium content that increases southwards. A comprehensive study of 5 m grid was carried out on an episyenite feature associated with this granite to delineate uranium mobilization and traps. Uranium and Thorium surface distribution studies (Fig. 3.5.5) revealed the

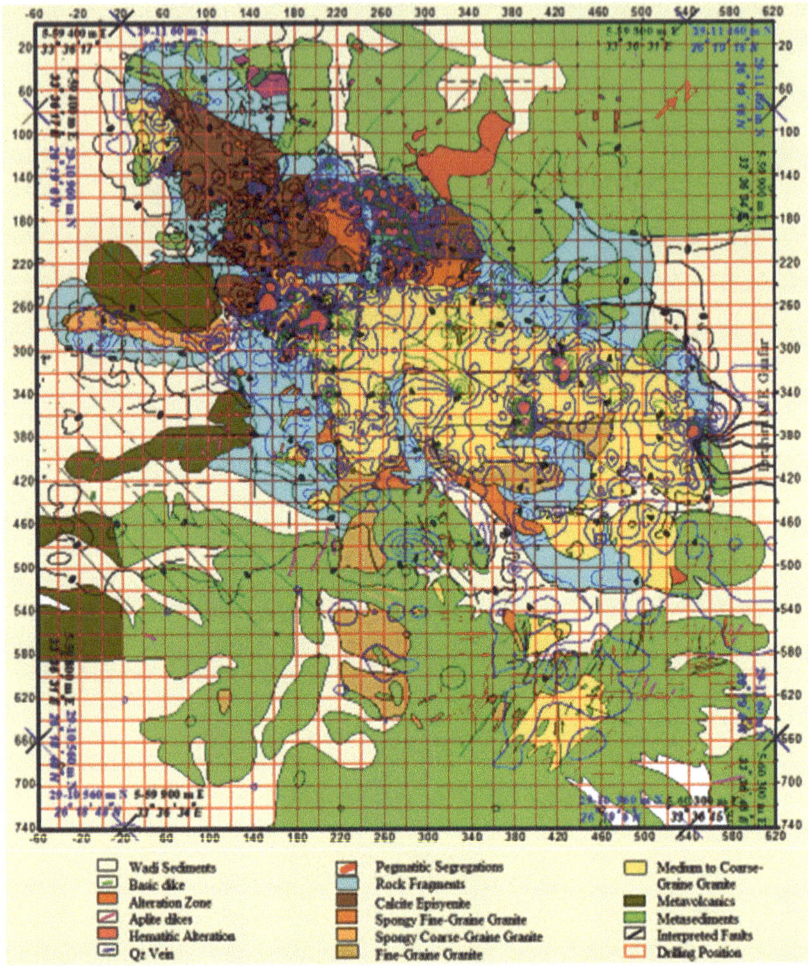

Fig. 3.5.5 ***eU-eTh/3.5 ratio contour map for the Episynite Zone, Southeastern Kab Amiri area, Central Eastern Desert, Egypt. Contour Interval = 2 Black contour lines ≤ 0 Blue contour lines = 2–8 Green contour lines = 10–16 Violet contour lines = 18–22 Red contour lines = 24–100.***

uranium mobilization trend. A structural and very low frequency geophysical mapping helped in locating structural traps and drilling targets. Drilling to a depth of 120 m showed the continuation of the anomalies. The well logs showed that the oxidation zone continues down to the present drilling depth of 120 m. Therefore, to get to the trap in the reduction zone where major deposits are anticipated, drilling must be continued down to a depth more than 300 m.

CHAPTER 3.6

Conclusions

Despite the technological advancement in new methods of surveying, the geological survey remains a cornerstone in the search for uranium ore deposits especially in selecting target areas and specific subsequent methods of surveying. The technological advancement in methods of surveying makes the process of uranium exploration a more reliable, economical, and accurate one. Usually, not all of the prospecting methods are used in one area, but rather some of them are tailored to be used according to a geological survey of the target area and reinterpreted data collected during previous studies as well. This was obvious in the methodology used in the case studies especially in Kab Amiri area where specific methods of surveying were used to identify highly potential areas. Subsequent drilling, well logging, and bulk sampling were used to determine the depth at which uranium deposits could be located.

References

Asfahani, J., Al-Henta, A.R., Aissaa, M., 2006. Identifying spectrometric signatures of phosphate deposits and enclosing sediments in Al-Awabed area, Northern Palmyrides, Central Syria, by the use of statistical factor analysis. Appl. Radiat. Isot. 64 (9), 1082–1090.

Birnie, R.W., Francica, J.R., 1981. Remote detection of geobotanical anomalies related to porphyry copper mineralization. Econ. Geol. 76, 637–647.

Department of Primary Industries, Australia, 2008. http://www.dpi.vic.gov.au/DPI/Vro/vrosite.nsf/pages/landform_glossary_radio. (Accessed August 9, 2021).

Dickson, B.L., 2004. Recent advances in aerial gamma-ray surveying. J. Environ. Radioact. 76 (1-2), 225–236.

Durrance, E.M., 1986. Radioactivity in Geology, Principles and Applications. Ellis Horwood Ltd., NY, p. 441.

Elkattan, E.M., Abdulhadi, Hassan M., Rabie, Said I., Hamdy, I.E, January, 1996. Hassanein application of ground geophysical data to uranium mineralization in the El-Missikat area, central Eastern Desert, Egypt. J.f Afr. Earth Sci. 22 (1), 81–91.

Geological survey of Canada, 2008. www.gsc.nrcan.gc.ca. (Accessed August 9, 2021).

Griffiths, D.H., King R.F., 1981, Applied Geophysics for Geologists and Engineers: The Elements of Geophysical Prospecting – Pergamon, Oxford, UK.

Hanna, W.F., 1969. Negative aeromagnetic anomalies over mineralized areas of the Boulder batholith. U.S. Geological Survey Professional, Montana, Paper 650-D, pp. 159–167.

Hoover, D.B., Grauch, V.J.S., Pitkin, J.A., Krohn, D., Pierce, H.A., 1991. Getchell Trend Airborne Geophysics-An Integrated Airborne Geophysical Study Along the Getchell Trend of Gold Deposits. North-Central Nevada.

Hoover, D.B., Douglas, P.K., David, C.C., 1995. Geophysical methods in explorationand mineral environmental investigations. http://pubs.usgs.gov/of/1995/ofr-95-0831/CHAP3.pdf. (Accessed August 9, 2021).

Knepper, D.H., et al., 1989. Mapping hydrothermal alteration with Landsat thematic mapper data. In: Lee, K. et al (Ed.), Mineral deposits of North America, volume 1, Remote sensing in exploration geology. American Geophysical Union, Field trips Guidebook: Washington, DC, pp. 13–21.

Melkov, V.G., 1956. Methods of uranium prospecting. Sov. J. At. Energy 1, 85–91.

New South Wales Minerals Council LTD. 2008. https://www.nswmining.com.au/. (Accessed August 9, 2021).

Palacky, G.J., 1986. Airborne Resistivity Mapping. Energy, Mines and Resources Canada, Ottawa, Canada, p. 195.

Palacky, G.J., 1988. Resistivity characteristics of geologic targets. In: Nabighian, M.N. (Ed.). Electromagnetic Methods in Applied Geophysics, volume 1. Society of Exploration Geophysicists, Tulsa, OK, pp. 52–129.

Palacky, G.J., 1993. Use of airborne electromagnetic methods for resource mapping. Adv. Space Res. 13 (11), 5–14.

Paterson, Norman, April 23-26, 1995. Application of geophysical methods to the detection and monitoring of acid mine drainage. In: Bell, R.S. (Ed.), Proc. Symposium on the Application of Geophysics to Engineering and Environmental Problems. Orlando, Florida. Environmental and Engineering Geophysical Society, pp. 181–189.

Raines, G.L., Lisk, R.E., Schafer, R.W., Wilkinson, W.H., eds., Geology and Ore Deposits of the Great Basin, Geological Society of Nevada, NV, vol. 2, pp. 739–758.

Rabie, S.I., Assaf, Hassan S., El Kattan, El Sayed M., Osman, Helmmy S., 1996. Uranium exploration in El-Missikat prospect area, Central Eastern Desert. Egypt Radiat. Phys. Chem. 47 (5), 769–774.

Sato, M., Moony, H.M., 1960. The electrochemical mechanism of sulfide self-potentials. Geophysics 25, 226–249.

Shumacher, D., 1999. Surface geochemical exploration for petroleum. In: Beaumont, A.D., Foster, N.F. (Eds.), Exploring for Oil and Gas Traps, Treatise of Petroleum Geology, Handbook of Petroleum Geology. AAPG, Tulsa, OK 18-1 to 18-27.

Sumner, J.S., 1976. Principles of Induced Polarization for Geophysical Exploration. Elsevier, Amsterdam, p. 277.

William Murray Telford, L. P. Geldart, Robert E., Sheriff (Contributor) L. P. Geldart, Robert E. Sheriff. Applied Geophysics, Cambridge University Press, Cambridge, 1990.

CHAPTER 4

Geosystems model of uranium deposits

CHAPTER 4.1

Introduction

Bodies of knowledge in the sciences pass through the descriptive stage, then the classification stage, then patterns are identified, and generalizations, models, and grand theories are made. Adaptations and revolutionary changes and reinterpretations can happen. This is the story of our knowledge of uranium deposits over the past 60 years. The movement of uranium through the crust and surface of the earth is of interest. There are low uranium concentration values called background which vary with bedrock type, and there are enrichments, concentrations, or positive anomalies of uranium concentrations. Some enrichments evolve into mineral deposits, the exploitation of which can bring great benefit and wealth to society. Uranium mineral deposits are the only source of all uranium used by society, mainly to produce nuclear energy. Uranium is a very versatile geochemical element, meaning that it can be enriched in many different geological environments. Thus there are many different types of uranium deposits, called "deposits types" (Fayek, 2013; IAEA, 2009). There are many uranium mines and occurrences throughout the world but they are not uniformly geographically distributed. Some of these mineral deposits have been well-studied, such that a model has been constructed synthesizing geological properties/characteristics from similar many deposits. A mineral deposit model interprets the origin of the deposit and is data rich, helpful in understanding processes, and is useful in mineral exploration (US Geological Survey, 2010).

The rock cycle contains boxes as geological entities and arrows between the boxes as geological processes. The geochemical cycle of a chemical element is the rock cycle expanded to consider the movement of that chemical through the detailed geological processes and reservoirs. Geochemical cycles or segments can be constructed to correlate with a specific interest at hand.

DOI: https://doi.org/10.1016/B978-0-323-90992-1.00004-X

An environmental or mineral resource cycle could be derived. To present the uranium mineral deposit types within the geochemical and geotectonic cycle of the earth processes is the objective of the present research. Understanding the position and characteristics of the mineral deposits in cycles is an objective of economic geology. Application of that knowledge in exploration geology can provide needed resources and create wealth.

When supervised artificial intelligence is applied to mineral deposit models, the proper flow and order within the cycle is essential. The proper understanding requires a useful classification that logically connects entities or deposits. This research will present the segment of the geochemical cycle of uranium, which contains most of the recognized uranium mineral deposit types, and it will put them in a logical, sequential, order related to their geotectonic setting. Such an idea has not been presented previously.

Let us consider the natural geochemistry of uranium, beginning with the classification of uranium deposits. The geochemical diversity of uranium produces of large number of uranium deposit types and also a large number of classifications of those deposits. Geology of the surface varies from country to country. The geology of Chile is different from that of Germany. Likewise, the geochemical endowment (how much of that chemical is present in mineral deposits) of countries varies, such that the uranium endowment varies dramatically enhancing the economics and politics of the element. Furthermore, the differing geology between countries results in their uranium resources being represented by different deposit types. The United States of America in the 1960s and 1970s dominated production from sandstone hosted deposits. Canada dominates production today from (Precambrian) unconformity-type deposits.

CHAPTER 4.2

History and examples of classifications

Uranium was not a valuable commodity until approximately 1940. Fifteen years of military use was followed by increasing use in electricity generation, beginning about 1953, which now, 2019, produces about 20% of the electricity in the United States of America, 55% in France, and rapidly increasing amounts in China. Uranium is the only commodity for which the United Nations has its own agency. The International Atomic Energy Agency provides information on all things nuclear and promotes and monitors nuclear activities around the earth. The Red Book is an annual IAEA document summarizing uranium resource global reports for each year. The classification of deposit types is complicated. Over the last 50 years many different classifications have evolved (e.g., Fayek, 2013; IAEA, 2009; US Geological Survey, 2010; Bowie, 1970; Dahlkamp, 1993; Plant et al., 1999; Cuney, 2009; Cuney et al., 2012). Most of these classifications are based on the nature of mineralization, type of host rock, and morphology of ore body. The International Atomic Energy Agency (IAEA), the authority of global nuclear knowledge, established its Classification of Uranium Ore Deposit Types based on total uranium production/reserves per deposit type during the Cold War, with little reporting or knowledge of Russian or Chinese uranium geology.

A genetic classification arranges the three rock types—igneous, sedimentary, and metamorphic—around a circle, representing the rock cycle (Fig. 4.2.1A; Cuney, 2009). On the left are magmatic activities and on the right is a sequence of activities from sedimentary to metamorphic, all with associated uranium deposit types. There is very little geological or tectonic information associated with the diagram. Cuney (2009) and Cuney and Kyser (2009), elaborated a genetic classification that contains small geological cross-sections of the most important uranium deposit types, conveniently, all on one diagram (Fig. 4.2.1B), but neglecting tectonic and geological elaboration. Another classification of Uranium Mineral Systems (mineral deposits) is based on fluids (Fig. 4.2.1C). The nature of mineralizing fluids (i.e., hydrothermal, meteoric, seawater, or basinal brines) and/or their transport mechanisms in metamorphic, igneous, and hydrothermal system is not well understood All of the above knowledge is valid and useful, however it may be improved by integration with tectonics and the uranium geochemical cycle. There is a need to provide a unified, geological, geochemical, classification scheme/genetic model that should emphasize the genetic architecture of the deposits. In the present work, we particularly focus on uranium Geosystems to provide a more robust model of uranium deposits which should take into account the role of both geochemical and tectonic environments in the genesis of uranium mineralization.

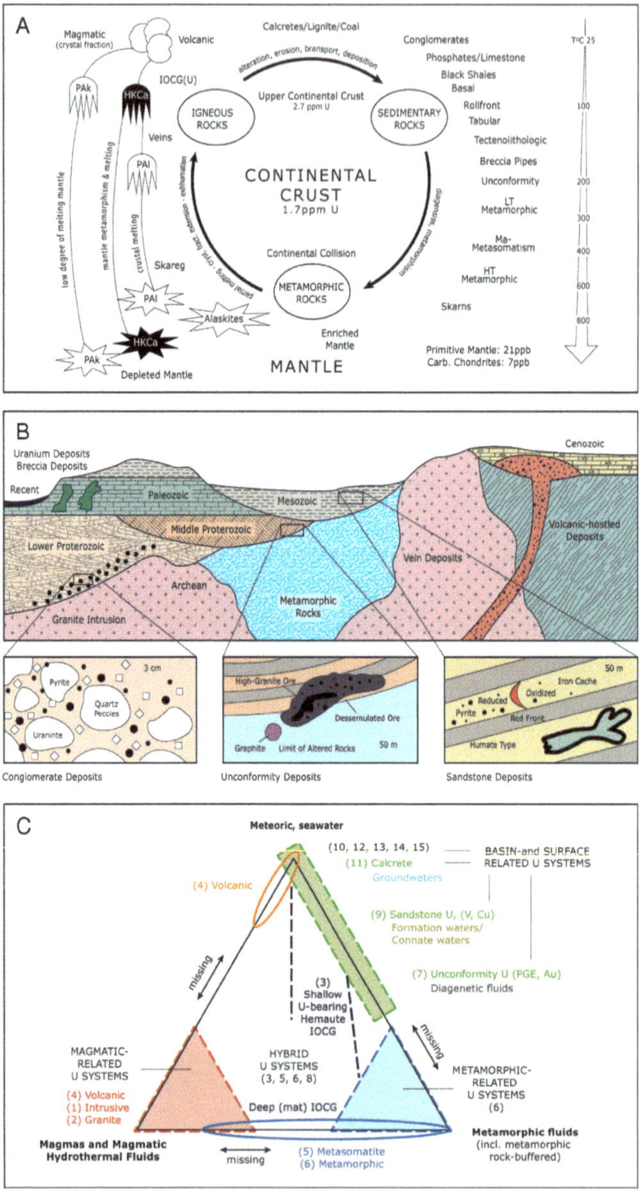

Fig. 4.2.1 ***Three depictions of classifications of uranium ore deposit types.*** (A) Uranium deposit cycle, called genetic classification scheme; Rock cycle in a circle, after Cuney (2009), and Fayek (2013). Three rock types in a circle are surrounded by 15 uranium deposit types. (A) Genetic classification (Cuney, 2009). Large geological cross-section with subsets of cross-sections of prominent uranium ore deposit types, all on one figure. (C) Triangular hydrogenetic classification of uranium deposits. End-member fluid compositions and types associated with depositional fluids of uranium ore deposit types (Skirrow et al., 2009).

CHAPTER 4.3

Geosystem model/classification of uranium ore deposits, guided by the geochemical cycle of uranium

The proposed Geosystem Model is given in Table 4.3.1, and it will be illustrated in a sequential series of crustal profiles with uranium deposit types. Historic uranium deposit types can be accommodated, and not eliminated. Equivalence to IAEA types is given in that table also. In Geosystem thinking, rift shoulders are very important as the starting point of the uranium geochemical cycle. Red, felsic, A-type, caldera-related, granites become the most important magmatic event (discussed later in the text). They and their volcanic equivalent are the source of the volcanic and part of the sedimentological component of uranium deposits. Several unifying themes are proposed. A brief summary follows. (1) The uranium geochemical cycle begins on rift shoulders with A-type granites. (2) Many felsic calderas are the surface expression of A-type granites. (3) The same intra-cratonic sedimentological basin is used through geological time to illustrate different sediment-hosted deposit types. (4) The surface of this same intra-cratonic basin is the site of multiple uranium deposit types that receive at least some of their uranium from ash-flow tuffs eruptions from felsic calderas (point 2 above) located on rift shoulders (point 1 above). (5) Outliers to the model exist.

Veins as a deposit type are not present as a subclass in the IAEA (2013) (IAEA, 2009) but were present in IAEA (2004). They are included here. Hydrothermal is here present as a Regime, #3, but consists of overlaps with magmatic and volcanic. Subclasses are intended to be geological features, environments, or deposits, but some process types of designations have been made, such as metasomatic and hydrothermal. Overlaps are natural. Awareness of them is emphasized. Certain relations are inherent in the geology. Magmatic evolves to metasomatic, which evolves to hydrothermal. Deposition/precipitation takes place from magmas at the start, and cooling causes aqueous fluids to be emitted (exsolved). Shapes may be dykes, irregular masses, etc., or veins, stratigraphic controlled deposition, etc. The understanding of the origin or genesis of a mineral deposit type requires the answers to the following three questions: (1) What is the geological source of the metals or materials? (2) What is the geologic/geochemical transport mechanism of the metal from the source to the enriched deposit? (3) What caused deposition of the metal? Potential general answers to (1) and (2) are offered at the end of Table 4.3.1.

Table 4.3.1 Geosystems classification. Uranium deposit types are here initially subdivided into Regimes, such as (1) Magmatic: M, (2) volcanic: V, (3) hydrothermal: H, (4) Platform: P, (5) intra-cratonic basin: B, (6) Other: O. Uranium sources (S) and transport mechanisms (T) are presented at the end. Numbers are given in the right-hand column, showing the correlation between Geosystems on the left and IAEA (2013) (IAEA, 2009) on the right.

Geosystems	IAEA class	Notes	Additional notes
Magmatic Fig. 4.3.3	iaea = 2013 IAEA		
1.1 Primary granitic differentiates	(1.22, 2 iaea)		
1.1.1. Widespread red, uranium rich granites			
1.1.2. Small, highly uranium-enriched differentiates			
1.2 Pegmatites	(2 iaea)		
1.3 Metasomatic	(5 iaea)		
1.4 Veins; and see hydrothermal			
1.5 Breccias			
1.6 Skarns (adjoining host rocks)	(2.2 iaea)		
1.7 Disseminations			
V2 Volcanic Fig. 4.3.4	(4 iaea)		
2.2. Ring dykes			
2.3. Resurgent domes			
2.4. Rhyolite flow domes			
2.5. Pyroclastic flows and breccias			
2.6. Mafic flow breccias			
2.7. Caldera boundary breccias			
2.8. Pre-resurgence intra-caldera volcaniclastic and moat facies			
2.9. Post-resurgence intra-caldera volcaniclastics and breccias			
2.10. Outflow facies step faults			
2.11. Altered basal vitrophyre			
2.12. Outflow facies lapilli tuffs, lithophysal zones			
2.13. Unwelded basal lithic (limestone) rich zones of aft			

Geosystems	IAEA class	Notes	Additional notes
2.14. Channels cut into basement of volcanics filled with conglomerate, sandstones, arkose, organic debris			
2.15. Volcaniclastic facies			
2.16. Organic-rich lacustrine facies			
2.17. Playa, calcrete	(11.3 iaea)		
2.18. Karsts and faults and structural features beneath volcanics	(13 iaea)		
2.19. Stratigraphic (clastic units) features beneath volcanics			
2.20. Chemical rock features beneath volcanics			
H3. Hydrothermal			
3.1 = 1.4,1.5, magmatic hydrothermal veins	(4, IAEA 2004)		
3.2 = 5.2, 5.4., meteoric and connate water (convecting groundwater cells)	(4. IAEA 2004)		
3.3 Other			
P4. Platform Fig. 4.3.5			
4.1 Collapsed breccia pipe	(8 iaea)		
4.2 Karst deposits in limestones	(13 iaea)		
4.3 Carbonaceous shale–related stockwork (Mo, Ni, V, As, Sb)	(15.2 iaea)		
B5. Intra-cratonic basin			
	5.1. Archean and Paleoproterozoic conglomerates, Fig. 4.3.6A, IAEA subclasses= 5.1.1 and 5.1.2. Witwatersrand, Pre-oxygen, pyrite+uraninite clasts.	(10 iaea)	

(Continued)

Table 4.3.1 *(Cont'd)*

Geosystems	IAEA class	Notes	Additional notes
	Water transports clasts; uranium in clasts not water;		
	5.2. Proterozoic unconformity, Fig. 4.3.6B, source water Penetrates deeply, graphite and other reductants. IAEA subclass:	(7 iaea)	
		5.2.1 Above the unconformity 5.2.2 At the unconformity 5.2.3 Below the unconformity	
	5.3. Post-Devonian Phanerozoic sandstones, Fig. 4.3.6C, Source water penetrates shallow to moderate depths to encounter organic material from plants, and/ or methane. IAEA subclasses:	(9, iaea)	
		5.3.1. Basal channel deposits, 5.3.2. Tabular deposits, 5.3.3. Roll-front deposits, 5.3.4. Tectonic-lithological deposits, 5.3.5. Mafic dykes/sills	

Geosystems	IAEA class	Notes	Additional notes
	5.4. Present basin surfaces and adjacent ocean platforms, Source: volcanic ash from calderas, Fig. 4.3.6D,		
		5.4.1. = 2.17, calcrete	
(surface and very shallow groundwater) (11.3 iaea)			
		5.4.2. coal, lignite, peat	(11.1, 12 iaea)
		5.4.3 black shale	(15.1 iaea)
		5.4.4 phosphorite	(14 iaea)
6. Other			
6.1 Rift, carbonatite, Fig. 4.3.2 (1.2.3 iaea)			
6.2 Convergent-subduction, porphyry copper, Fig. 4.3.7A (1.2.1 iaea)			
6.3 Convergent-collision, anatexis, metasomatism, metamorphism, Fig. 4.3.7A (1.1, 5, 6 iaea)			
6.4 Strike slip—regional fault family, Fig. 4.3.7B			
6.5 Strike slip—pull apart (Olympic Dam), Fig. 4.3.7B (3 iaea)			
Potential uranium sources			
S1. Large volume felsic intrusives, red granites, beneath calderas, and in batholithic masses.			
S2. I Lightly differentiated tops of magma chambers beneath calderas may produce uraniferous granites (deposit type MI .1.) and host rocks, metasomatic and hydrothermal phases			

(Continued)

Table 4.3.1 *(Cont'd)*

Geosystems	IAEA class	Notes	Additional notes
S3. Large volume felsic ash-flow tuffs, erupted from Sl and S2 magmas and calderas.			
S4. Resistates (monazite) in sediments			
Potential uranium transport mechanisms			
T1 Magma movement			
T2 Metasomatic			
T3 Hydrothermal			
T4 Erupted ash			
T5 Distal erupted ash			
T6 Surface water			
T7 Groundwater			
T8 Ocean water			

Answers to vary with deposit type. We do not have all of the answers for most deposit types, but knowledge is summarized (Fayek, 2013). Degrees of knowledge vary with deposit type. The present study does not maintain that it is all inclusive. The goal here is, for a large number of uranium deposit types, to present them integrated into the normal flow of the geochemical and tectonic cycle. In the proposed Geosystem Model/Classification, no IAEA subtypes or classes are eliminated, but they may be grouped or affiliated with others having common geological characteristics. The objective is to understand the logical sequence of uranium deposit-type locations and events from igneous to sedimentary environments. This progression of the geochemical cycle of uranium, in many diverse geosystems, is the base of the genetic model/classification of uranium deposit types presented in this work (Table 4.3.1). There are several outlier deposit types, which are classified as others or outliers (Table 4.3.1).

In the proposed Geosystem model uranium deposit types are separated into six regimes or general types. The normal flow of the geochemical cycle of many elements begins at the magmatic stage and moves on to sedimentological sinks. Uranium as well as thorium are relatively enriched in felsic igneous rocks, with their concentrations in the average granite being 3 and 12 ppm, respectively. Both have a +4 oxidation state in magmas and behave in a similar fashion. The thorium to uranium ratio as observed in different magmatic rocks is an important geochemical tool (discussed later in the text). Quantitative data are not presented, but general areas on a diagram will document different uranium deposits. Weathering and oxidation change U+4 to U+6, the latter which is soluble in water, thus uranium and thorium get separated during weathering at the surface of the earth. Thorium-rich minerals, oxides, and phosphates have small amounts of uranium, are resistates, meaning they are resistant to chemical attack during weathering. They are concentrated by physical processes, such as into black sands. The versatile uranium proceeds on with a life of its own, and a variety of deposit types in which it can concentrate. To complete this introductory discussion, a unique feature of the earth system under consideration here is the volcanic eruptions by which thousands of cubic kilometers of ash are ejected with low but significant and soluble uranium. This is the uranium source in the ash-leach theory for the origin of uranium deposits. Airborne, ashborne, glassborne, uranium transport is significant.

Magmatic uranium deposit types

The geochemical cycle of uranium begins at continental rift shoulders (Fig. 4.3.1A). Igneous-hosted uranium ore deposit types most commonly occur in extensional cratonic settings. Mantle thermal upwelling melts granitic, cratonic crust and results in emplacing intrusive rocks and hosting major volcanic activities that have participation with uranium geochemistry.

Asthenospheric upwelling initiates the process accompanied by uplift of a welt of the overlying continent. The brittle continent breaks and that is the incipient rift. Strong extensional forces exist, tending to pull a continent apart. Prior zones of weakness within the crust exist, and break-up often takes place along them. Breaks take place in a near-vertical plane; Fig. 4.3.1B shows the eastern boundary of the Rio Grande rift in the central New Mexico (West et al., 2004). Such vertical breaks make the region amenable to lateral faulting (or maybe it was the other way around) and lateral displacement within rifts is common. Hot, asthenospheric magmas rise as thin vertical dykes into the felsic subcontinental lithospheric mantle. They pond in large sill-like bodies, and these could be stacked on the top of each other. This introduces much heat into the lower crust. Rising curved isotherms cause melting and anatexis at the base of the granitic crust, as shown in Fig. 4.3.1A.

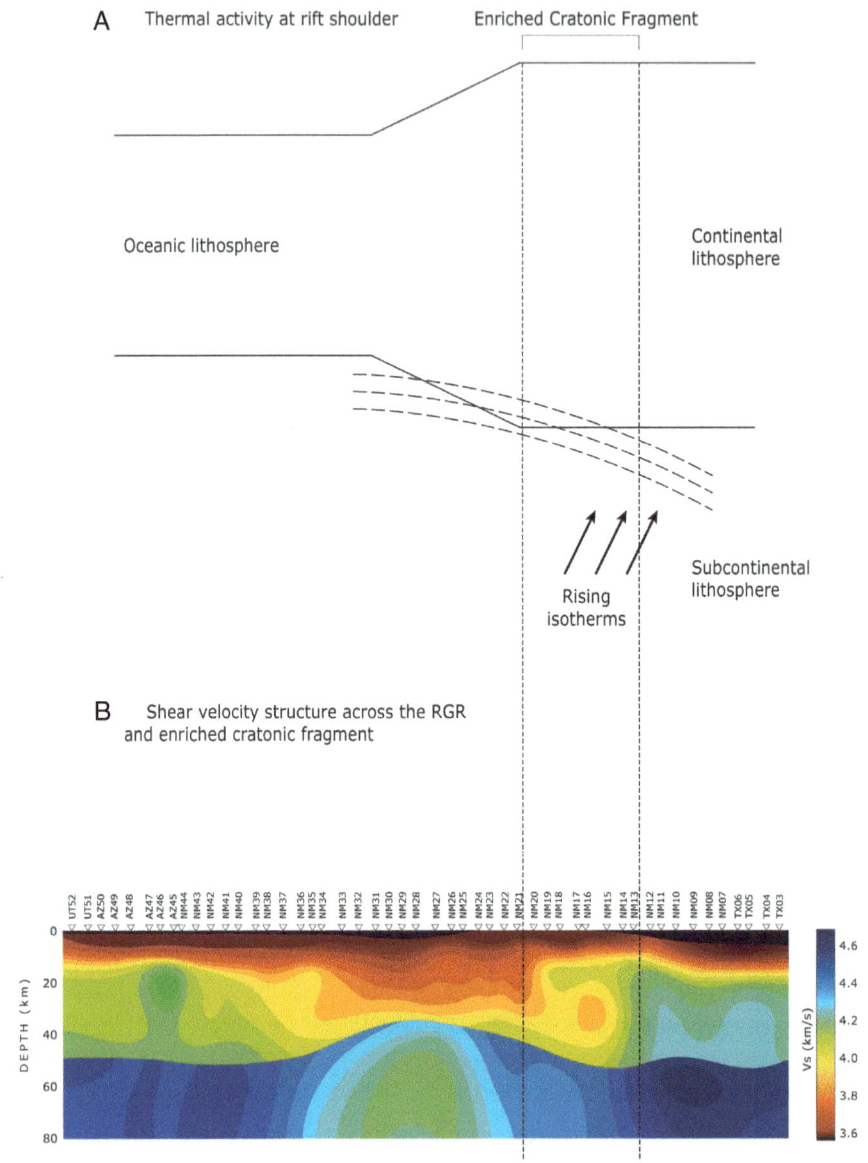

Fig. 4.3.1 ***Rift Shoulders–The start of uranium geochemical cycle.*** (A) Schematic crust/mantle vertical cross-section of a craton versus active rift, with rising isotherms, (B) La Ristra cross-section. This image was produced by measuring Rayleigh wave phase velocities for periods ranging from 7 to 150 s, the velocities were inverted to determine shear velocity structure using constraints on crustal structure provided by receiver functions, and constraints on mantle structure provided by teleseismic tomography. Image shows a symmetrical zone of low velocity centered on the Rio Grande rift (West et al., 2004).

These felsic magmas eventually move to ponding areas in the shallower crust. The felsic magmas constitute A-type igneous rocks, and together with the mafic asthenospheric material they constitute bimodal magmatism. The A-type designation is accompanied by numerous characteristics, including that they are F dominated, not Cl as at convergent plate margins, and they are enriched in U, Th, F, and Mo and rare earth element (REEs), and others. In the sub-caldera magma chambers, peralkaline magmas may undergo a high degree of magmatic fractionation (Figs. 4.3.2 and 4.3.3) and evolve to have a high gas content. It is proposed that a zonation exists from porphyry Mo systems at depth, to zones of enrichment in F, Be, U, heavy rare-earth elements (HREE), and others, with topaz rhyolite at the surface. In the present study, this is one of the most important environments/deposit types. It constitutes the intrusive portion of caldera magmatism on rift shoulders.

Fig. 4.3.3 documents the multiple subclasses within this environment and connects with Fig. 4.3.4 of the volcanic deposits. Fig. 4.3.4 can be

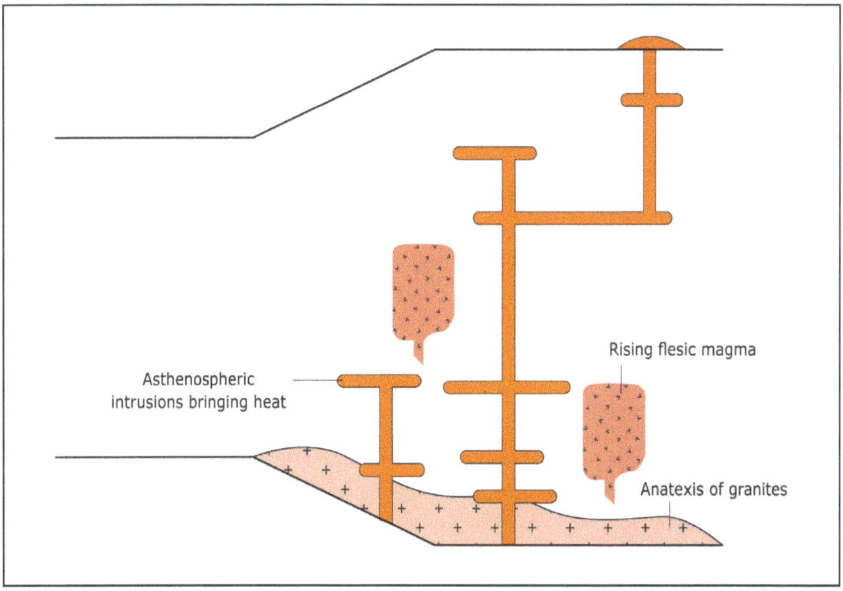

Fig. 4.3.2 ***Schematic crust/mantle vertical cross-section of a craton versus active rift, a result of rising heat in 2A.*** The felsic lower granitic crust is melting and producing magmatic blobs which rise toward the surface. Contemporaneously, dyke-like mafic intrusions occur and bring heat upwards. Their arrival near felsic blobs stranded in their assent may trigger upward movement of the blobs. In the intermediate crust the felsics may pond and crystalize into bodies of batholithic dimensions. Stocks sent off the tops of these magmatic masses may be the magma masses underneath a caldera.

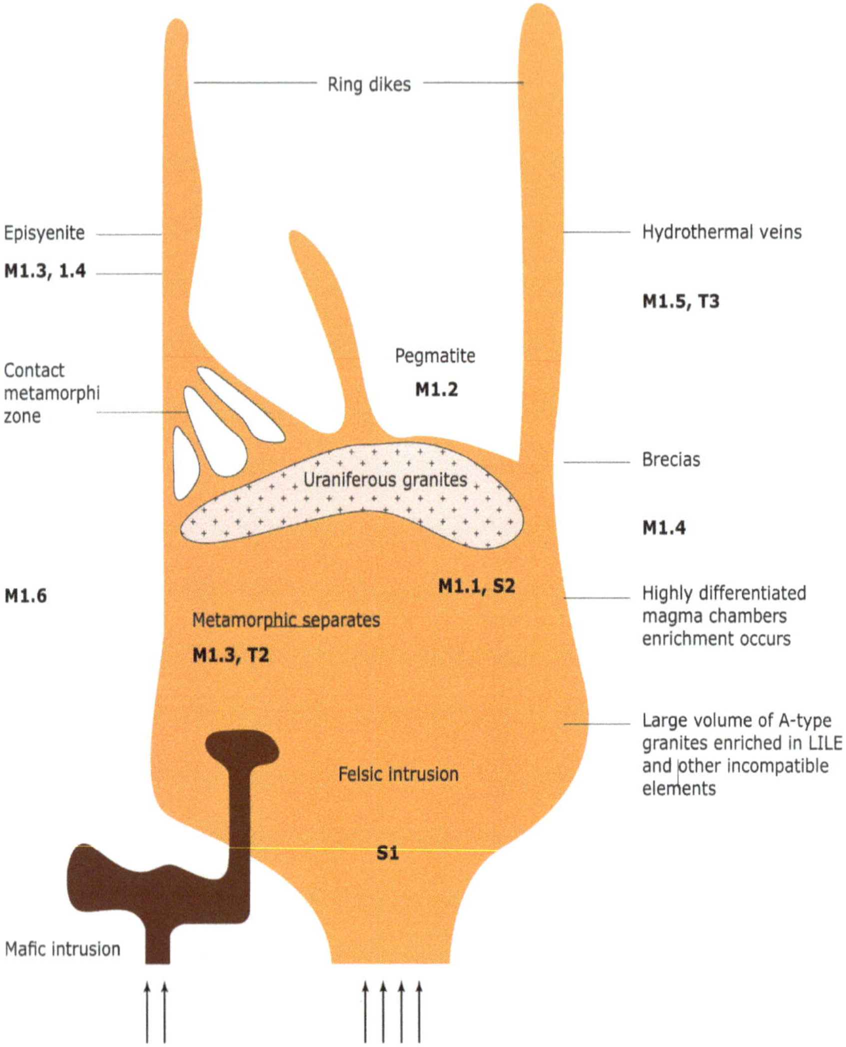

Fig. 4.3.3 ***Rift shoulder located large felsic magma chamber underlying a caldera.*** Schematic crust/mantle vertical cross-section of a craton versus active rift, a result a rising heat in Fig. 4.3.1A. Potential areas of uranium enrichment, some resulting occurrences and known ore deposits and/or regions which are being mined, are noted by number. M: magmatic, S: source, T: transport. Caldera magmatism gives evidence of repeated magmatic replenishment. The arrival of these masses at the surface initiates the sequence of events known as the evolution of caldera magmatism, which may repeat itself, and overlap with others. And those mafic dykes may also find their way upward, characterizing bimodal magmatism. Mafic extrusive events in calderas are known. This diagram extends up to the pre-volcanic surface. Metasomatic, pegmatitic, breccias, contact metamorphic, and hydrothermal zones are found in multiple areas in this figure.

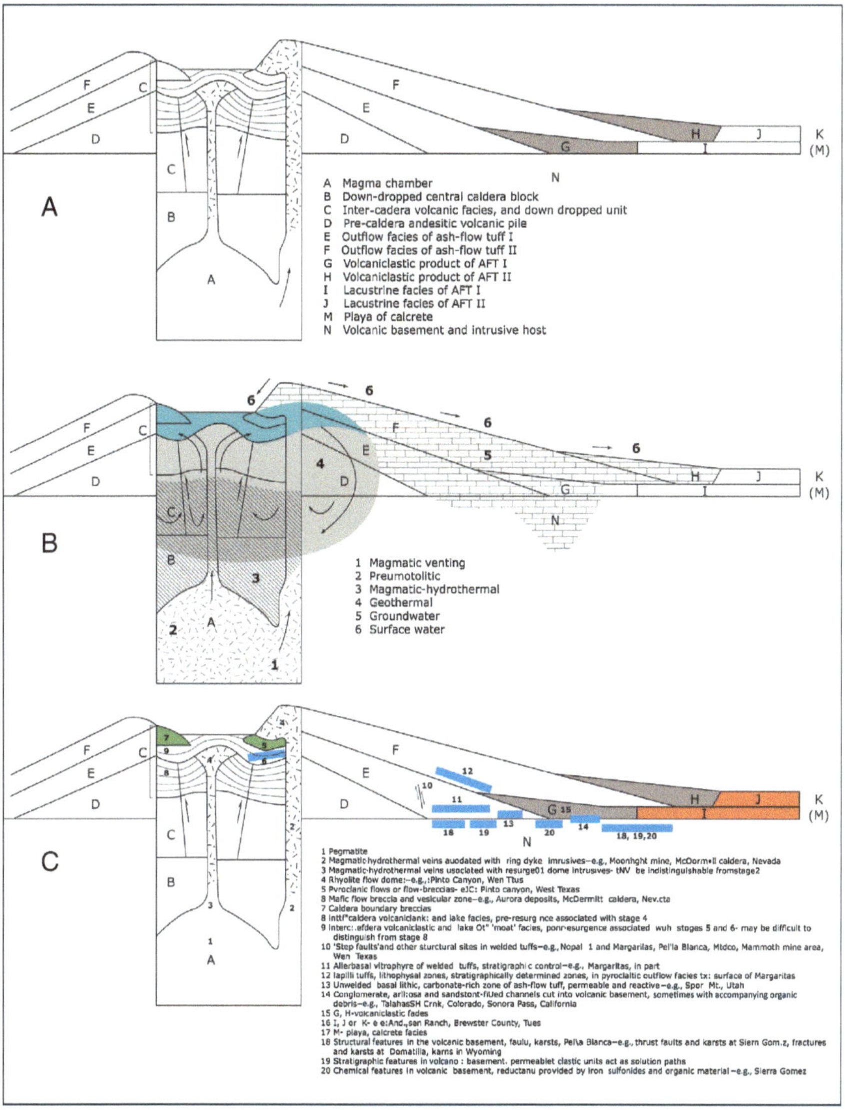

Fig. 4.3.4 ***On top of the felsic magma chamber depicted is the magmatic regime, analogous to the Valles Caldera in New Mexico.*** Calderas have an ideal sequence of events corresponding to structural, igneous, and hydrothermal circumstances; however, the progression can be interrupted, restarted, recycled, or continue happening nearly. Fig. 4.3.4 uses the same template of inner and outer caldera facies (see legend) to illustrate the ideal volcanic (intrusive) edifice, 4A and 4B show the regions of hydrothermal activity; 4C shows the locations of known uranium occurrences, some being productive mines (Goodell, 1985).

a caldera above an intrusive of batholithic size, generally an A-type red granite. An extensive differentiation sequence is recognized, and at least one stage is associated with mineralization. Primary, intrusive, fluorite can be present. The tops of the highly differentiated magma chambers may be the sites of formation of endogranite uranium deposit types 2.1, and of pegmatite zones. Perigranite deposit types 2.2 would occur in the adjacent host rock. Ex-solution or separation of an aqueous metasomatic/hydrothermal phase often takes place. Magmatic differentiates and aqueous fluids move upward causing breccia zones, dykes, and shallow intrusive masses, in many different sites, however just one is indicated in Fig. 4.3.3. Some of the intrusives may be enriched by differentiation, and the aqueous fluids even more so, all variously enriched in Li, Be, Sn, Ta, HREEs, and U, from fluorine-dominated systems.

Fig. 4.3.3 can also be zoned from porphyry Mo deposits at depth to laccoliths and extrusives, specifically topaz rhyolite, a shift may take place from LCT chemistry (lithium-cesium-tantalum) to NYF (niobium-yttrium-fluorine) chemistry, and this late stage metasomatic behavior is strongly associated with the deposition of HREEs. Mineralized skarns also take place in the host rocks.

Widespread, red, uranium-enriched granites constitute the first entry into the Geosystems Classification, type 1.1.1. Primary granitic differentiates, it constitutes a better uranium source, S1, than deposit; it appears in Fig. 4.3.3, and could be large in dimensions. A most highly differentiated, moderate volume, cap, or upper zone to the magma chamber may form which constitutes a potential uranium deposit, as demonstrated, perhaps, by the Rossing deposit, in Namibia, type 1.1.2. Several other uranium deposit types may occur nearby, in Fig. 4.3.3. Late stage magmatic-metasomatic activity produces pegmatites (type 1.2, 1.3), often in dykes or the host rock.

Metasomatic-hydrothermal activity produces deposits in host rocks as skarns (type 1.6) and breccias, veins, and pipes to the surface (types 1.4, 1.5). There are multiple geological opportunities for multiple deposit types to be present in Fig. 4.3.3. Magmatic uraninite present as an accessory mineral is optimal although it is quite rare. The igneous source rocks for the Witwatersrand Archean uranium deposit, Type 5.1, had magmatic uraninite. Magmatic uraninite is present in the Rossing mine, Namibia, and at the Bokan Mountain deposit in Alaska. Rossing actually resides in a dislodged fragment of stable block out in a mobile belt and is characterized by alaskite produced at anatexis or early melting during a thermal event in the mobile belt. Magmatic-hosted uranium enrichments, type 1.1.2, are best illustrated by Bokan Mountain, Alaska. This deposit resides tectonically in a long narrow terrane as a part of an accretionary assembly in southeast Alaska. Bokan Mountain does give a more reliable understanding of a high-level magma chamber and massive dykes, but leaves the question open at times as to the difference between magmatic and hydrothermal. Its relationship to any magmatic province is likewise not known.

Volcanic uranium deposit type regime

Figs. 4.3.2 and 4.3.3 show a schematic vertical cross-section through a rift shoulder felsic magma chamber, and Fig. 4.3.4 is the complex volcanic edifice above it. Fig. 4.3.4A shows simplified geological events of caldera evolution. This is then used as the base or template for events in the hydrological system (Fig. 4.3.4B) and the occurrence of known uranium deposits (Fig. 4.3.4) overlaps the magma chamber and adds caldera events plus outflow facies cooling units plus volcaniclastic stages. Sites 2 and 3 from Fig. 4.3.4 legend are hydrothermal parts of Geosystems.

An upward moving magma takes place by stress buildup, stress release, fracture/fault formation of intrusive dykes, formation of large xenoliths in the upper magma chamber, and eventually assimilation. Such influences in chemistry may be less than those presented by a highly differentiating, F dominated, metasomatic to hydrothermal fluid, produced from a differentiating magma. It is suggested that there is a porphyry Mo at depth, and a topaz rhyolite at the top. In between there can also be Be resources in skarn mineralization and ash-flow tuffs. Felsic calderas are not the only surface expression of these systems. In West Texas, a ring of HREE-enriched rhyolite laccoliths overlies a presumed Mo porphyry deposit. Approaching the surface, the magmatism may form ring of dykes, rings of laccoliths, rings of eruptive centers, or other features. Radial dykes and eruptive centers are also known. Major ash-flow tuff eruptions are of gas, magma, crystals (phenocrysts), and rock fragments. The magma chills quickly to glass and phenocrysts. Uranium is in the glass, and the glass is quickly destroyed to clay in surface weathering, and at the "first rainfall" much uranium is released into the hydrological system. Volumes of ash erupted from one event have reached 10,000 cubic kilometers, with 1000 cubic kilometers being common. This concept is staggering. Furthermore, calderas occur in groups, overlapping, concentric, nearby but connected by grabens. Caldera volcanism occurring on cratonic rift shoulders has the proper chemistry to be relevant to uranium deposits. Topaz rhyolites are one volcanic manifestation of potentially fertile systems. Collapse accompanies eruptions, producing different, intra-caldera environments consisting of breccias, lake sediments, mafic flows, etc., in Fig. 4.3.4. Individual caldera histories can be complex. This brief summary serves as a guide to Fig. 4.3.3 which depicts the intrusive portion of the system and Fig. 4.3.4 which depicts the extrusive portion. Calderas often occur in clusters, further complicating the situation.

During a volcanic eruption, chemistry of accessory minerals, the amount of liquid magma to become quenched glass at the time of eruption, and trace element chemistry are very important. What is uranium doing during all this? It will get partitioned into the magma rather than phenocrysts. Upon eruption, the magma is quenched to glass. Uranium is mainly in the glass. Deposited as an ash-flow tuff, the glass is unstable. Uranium is then released as the glass alters to clay, and the uranium can go on to form ore deposits. If uranium gets trapped within zircon as an accessory mineral, it is not as readily available for later reprecipitation to form ore deposits, because zircon is inert, a resistate. Magmatic uraninite present as an accessory mineral would be a slightly different scenario, although it is quite rare. The igneous source rocks for the Witwatersrand Archean uranium deposit, Type 5.1, had magmatic uraninite. Magmatic uraninite is present in the Rossing mine, Namibia, and at the Bokan Mountain deposit in Alaska. Caldera volcanoes are a common type of surface expression of extension. The Basin and Range Province of western North America is a modern-day example of such regional extension. Volcanic rifted margins are characterized and recognized by the bimodal chemistry, mafic and felsic, of volcanism at rifted margins. The silica-rich component is the environment of uranium enrichment and deposits. Examples are the West Texas Igneous Province, the Mogollon/Datil Volcanic Province, and the San Juan Volcanism, just around the southwest of the United States of America. Some of these calderas are part of the family known as Super volcanoes, which are exceedingly explosive and large. Some felsic igneous provinces have significant uranium (West Texas Igneous Province), whereas others do not (Datil-Mogollon). Subcontinental lithosphere geochemistry and thermal history are important.

Deposit type 1.6 consists of the contact metamorphic environment, adjacent to magmas enriched in uranium. Certainly, a magma rich in fluorine might have hydrothermal fluids efficient at limestone replacement and removal. This realistic deposit type, however, has but few potential examples. The Mary Kathleen deposit in Australia may be of this type. Within the sub-caldera magmatic, intrusive-related, environments, are hydrothermal veins and breccias. These may take many different forms and will transition upwards into hydrothermal veins and breccias of the Volcanogenic region.

Hydrothermal

In the Geosystems Model, there is substantial overlap between adjacent regimes, as seen in Figs. 4.3.3 and 4.3.4. The Hydrothermal Regime is an example. It characterizes Magmatic types 1.4 through 1.6, and Volcanic types. Fig. 4.3.4B shows schematically the regions of influence of various waters, including metasomatic/pneumatolytic, magmatic hydrothermal, geothermal, groundwater, surface water, as presented in Fig. 4.3.1. Veins were a uranium deposit type in prior versions of the IAEA classifications; it is retained here to be inclusive and recognize the overlapping of deposit types.

Platforms

The intrusive #1 and volcanic #2 regimes of uranium deposit types (Table 4.3.1), discussed above, have significant uranium deposits themselves; however, they are also thought of as source areas for uranium to now be transported largely as a part of the hydrological cycle, but also significantly, by clastic material and airborne ash. Igneous rocks and minerals are the major sources of uranium to the geochemical cycle, and large sedimentary basins in intra-cratonic or passive margin environments are the major depositional reservoirs to which that uranium is headed. Between these regions, the source of uranium and the ultimate depositional site or reservoir, there frequently is an intervening region here called a platform (Fig. 4.3.5). This is not the subaqueous platform-flooded margins of the continental environment; it is an above water plateau or platform able to receive felsic volcanic ash deposits (Fig. 4.3.5). The Colorado Plateau in the United States of America is the type example of Geosystems Deposit Type 4, the Platform. Numerous IAEA uranium deposit types are logically grouped here.

Geosystems Model deposit type 4.1 is IAEA 8, collapsed breccia pipes (Wenrich, 1985) which are found in abundance in the United States of America in thick Mesozoic limestones and dolomites, are small high-grade deposits in extensive vertical breccia zones, and which have been deciphered to have been collapsed. Ash-leach processes are believed to have

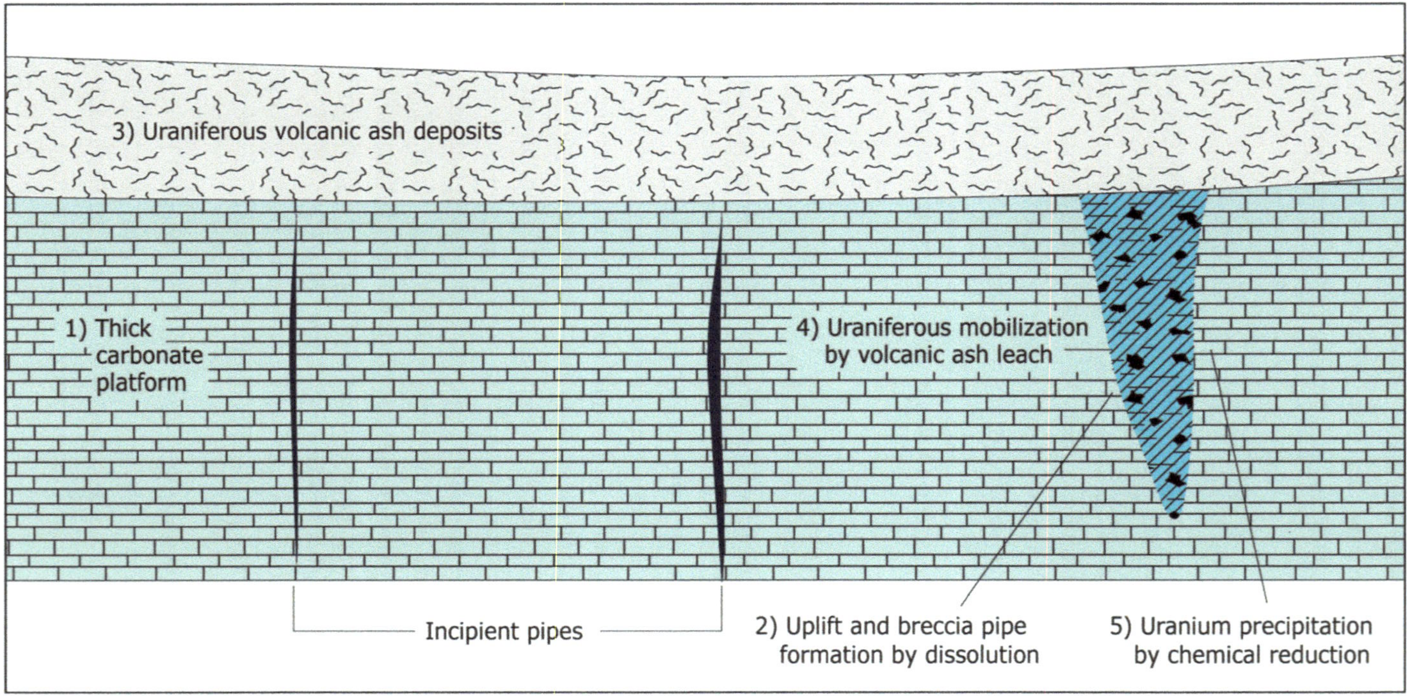

Fig. 4.3.5 ***The conditions necessary for the Platform group of uranium deposit types.*** Thick sequences of limestone and associated rocks are deposited widespread on submerged platforms in Paleozoic and Mesozoic time. Upon emergence, dissolution is initiated; forming small karsts, and evolving with stratigraphic control into lateral karst systems, and evolving with structural control into carbonate-hosted breccia pipes. The figure emphasizes the latter. Uranium arrives in later volcanic ash deposits. Uranium from the abundant glass moves into the groundwater, and hydrologically is channeled down breccia pipes.

precipitated the uranium in pyritic zones. The uranium deposits contain from 1 to 5 M tons uranium, and are relatively high grade at 2.0%. There are several thousand potential pipes in northern Arizona and possibly elsewhere in the world. The discovery and development of these deposits required the creation of this new uranium deposit type. Other similar deposits, but more karst like, exist in the Madison limestone in Montana, and in China. An alternative classification for this deposit is Karst zones in limestones (13 iaea), which is similar, however with a more limited vertical extent. Carbonaceous shale–related stockwork (15.2 iaea) may result from the same karst/collapsed pipe processes except that the host rock was carbonaceous shale. The geochemistry is similar between the pipe and the stockwork. For the above reasons, several IAEA uranium deposit types attributed to completely different deposit types are now, here, grouped together in the Geosystems.

Classification on the 3.4.PLATFORM is as follows:

3.4.1 Collapsed breccia pipe (8 iaea)

3.4.2 Karst deposits in limestones or dolomite (13 iaea)

3.4.3 Carbonaceous shale related stockwork (Mo, Ni, V, As, Sb) (15.2 iaea)

3.4.4 Fetid limestone facies (15 iaea, 2004)

3.4.5 Organic rich channels in dolomite or limestone (6.3 iaea)

Reaching out, in prior classifications, there was Fetid facies = IAEA (2004) to accommodate observations made on the Todilto limestone at Grants, New Mexico. It is included above as Platform 3.4.4, Fetid facies variation of limestone, which includes organic-rich facies, and rarely a reduced sulfidic-rich facies, fetid. This latter is the notable Jurassic, Todilto limestone, the enigmatic discovery outcrop of the Grants uranium District. The sulfide component of the rock served as a reductant for the uranium. Prior classifications also included 6.3 organic rich channels and karst in dolomite. They are placed under Platform type deposits here. These deposits are illustrated in the Um Bugma formation in the Sinai in Egypt (El Kattan and El Aassy, 1992).

Sedimentological branch of uranium deposit types intra-cratonic sedimentary basins and their surfaces

A preview of the forthcoming ideas is given next, and certain generalities applicable to much of the subsequent text are presented. An important generality is the introduction of a template of a vertical profile of an intra-cratonic sedimentary basin and its surface, which will be used repeatedly.

Basinal concepts here also apply to prograding, quiet, continental margins. It is a benefit for the conceptualization of uranium deposit geology to group various deposit types together, within the intra-cratonic sedimentary basins and their surface. This template is partly the line that separates Precambrian basement below from younger sedimentation above, the unconformity, and the template will be used as the starting point for generalized geological simulations of basins and the character of uranium deposits through time. The template will provide an understanding of the potential relationships between different uranium deposit types. The template also has a horizontal, youngest surface, which will be of great importance in understanding relationships between several additional uranium deposit types.

As discussed above, uranium is released into surface and groundwater by the chemical destruction or alteration of glass during weathering. Uranium is sourced or enters this system through surface and groundwater action, (1) S3, ash-flow tuffs by air fall, (2) S3, leaching of uranium bearing tuffaceous and glassy material, (3) S2, weathering of uranium bearing and fertile granites, (4) S4, diagenetic release from semi-resistate minerals (monazite) after deposition of the sediments, (5) leaching from the Precambrian basement. Numbers 4 and 5 above should not be underestimated. Number 4 means that some uranium enters this system as a minor element in solid solution in chemically and physically resistant minerals (resistate), such as zircon, monazite, and others. Only under unusual conditions can this uranium become available to a mineral deposit forming geochemical system.

For the most part, uranium is introduced into this basinal system dissolved in an aqueous fluid, and this input is from the igneous branch, discussed above. This dissolved uranium is in the +6-oxidation state, or valence, and it must usually be chemically reduced to the +4 state for precipitation of uranium minerals to happen, and for a uranium deposit to form. Precipitation of hexavalent uranium as a primary mineralogical species is not common. Reduction can take place by several pathways, but commonly takes place from the presence of organic matter. Thus, for continental sandstones, organic material is provided by plant life, and coal, lignite, and peat, are examples, and these are identified as uranium deposit types. Land plants are the result of an evolutionary development in life, and this evolution did not happen until the Devonian.

Pre-Devonian continental sandstones would have a hard time developing sandstone uranium deposits because of the lack of organic material. Following this same reasoning, the earlier Proterozoic intra-continental sandstone basins, which contain the economically most significant type of uranium target, have different types of uranium deposits than Phanerozoic or Post Devonian sandstones. One other large-scale global geochemical event was significant for uranium geochemistry and deposits, and this is the acquisition of increasing amounts of oxygen in the atmosphere, a little more than 2 Ga. With sufficient oxygen in the atmosphere uranium, minerals exposed during weathering get oxidized, and hexavalent +6 is mobile in water. Once this happened, uranium can get mobilized and then re-concentrated. Before this happened, weathering took place under a chemically more reduced atmosphere and uraninite was not oxidized but behaved as a resistate and heavy mineral in black sands. The above discussion demonstrates that intra-cratonic sedimentary basins have time-dependent changes in the behavior of uranium, and consequently, in the uranium deposit types which those basins host. This gives the organizational construction for the sedimentary basin hosted uranium deposits. It is envisioned that the Witwatersrand Archean quartz pebble conglomerates type 5.1 in Fig. 4.3.6A, Proterozoic unconformity type 5.2 in Fig. 4.3.6B, and sandstone type 5.3 in Fig. 4.3.6C were deposited in basins that had general similarity, as illustrated by the template. Over geological time, uranium was deposited in different places in the template, that same generalized basin. The basinal surface (Fig. 4.3.6D) is open to the atmosphere and thus also has the opportunity to source uranium from felsic ash deposits, S3, that is, to provide the uranium (and REE) to several known uranium deposit types, all of type 5.4. This novel, and generally neglected, transport mechanism is significant. Fig. 4.3.6D may get buried during subsequent history of the basin, at which point it would be susceptible to groundwater remobilization.

Geosystem deposit types in the sedimentological branch of uranium deposit types within intra-cratonic sedimentary basins and their surfaces are as follows:

3.5.1 Archean quartz pebble conglomerate (10 iaea)
3.5.2 Proterozoic unconformity types (7 iaea)
3.5.3 Phanerozoic (post Devonian) sandstone type (9 iaea)
3.5.4 basinal surface (11.1, 11.3, 12,14,15.1 iaea)

These geological time designations are generalized, and the details will not be discussed here. At depth and the oldest, with respect to uranium deposit type, type 5.1, (Fig. 4.3.6A) the Archean Quartz Pebble Conglomerate, IAEA 10 is a unique situation. The two IAEA subclasses can be accommodated by subdividing type 5.1 into 5.1.1 and 5.1.2. This deposit type formed in the Archean, so far back in the earth evolution

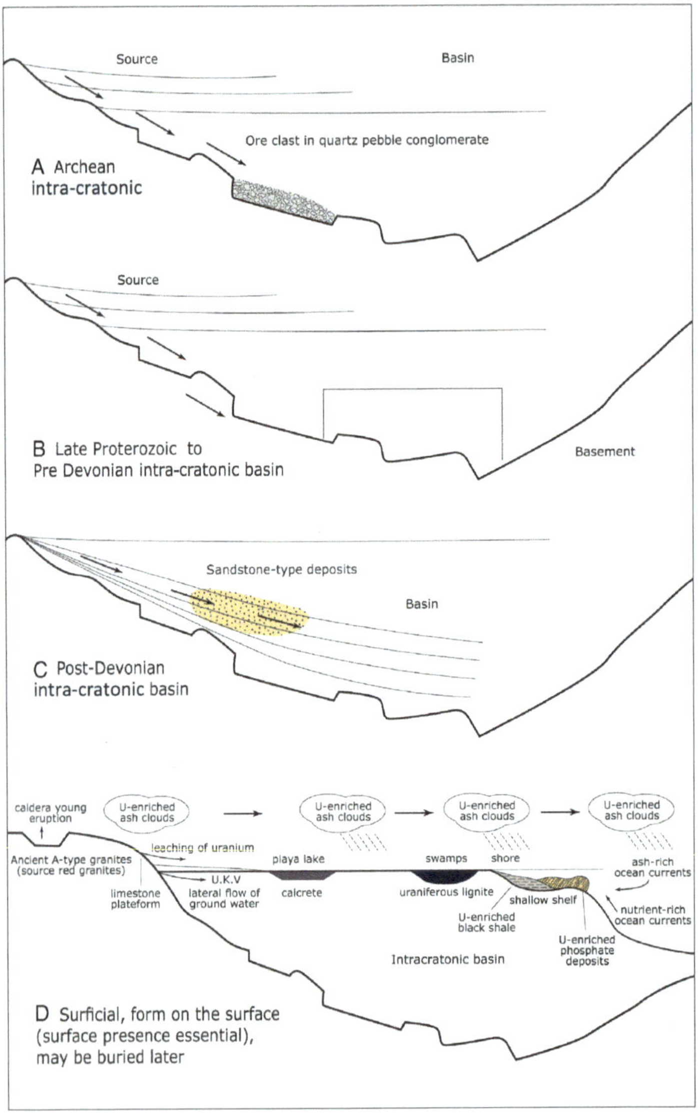

Fig. 4.3.6 ***Vertical cross-section of intra-cratonic basin.*** The same template of the basin is used for multiple images at different geological times. It consists of the base of the basin, the unconformity. (A) Archean sedimentation into large Witwatersrand basin in South Africa. Clasts include quartz pebbles, pyrite, uraninite, and gold. Physical transport of uranium as clasts. Non-oxygenated atmosphere prevents oxidation of pyrite and uraninite. (B) Proterozoic sedimentation into Athabasca basin, Canada, with uranium deposition near basal unconformity. Uranium source from adjacent uplands (S3) or monazite in clastic sediments (S4). Uranium transport in aqueous medium. (C) Phanerozoic, post Devonian, sandstone hosted, water transport, precipitation by reduction from organic material in sediments. (D) Surficial, form on the surface (surface presence essential), may be buried later.

that the chemistry of the atmosphere was different from today, such that there was no oxygen in the atmosphere. Thus, during the Archean minerals, which get oxidized and destroyed at the surface of the earth today, were stable during weathering. Thus, at the Witwatersrand in South Africa detrital pyrite, uraninite, and gold are found together in an Archean placer deposit. This deposit type is of limited importance in uranium production because Archean rocks are quite rare. There is one other place in the world besides the Witwatersrand, where this deposit type existed in commercial quantities, and that is Elliott Lake in Canada.

Before the Devonian and after oxygenation of the atmosphere, notably in the Proterozoic, there occurred a most important of uranium deposit types, the Proterozoic, intra-cratonic, basinal Unconformity Type. These are formed by (1) water leaching A-type granites or glasses (uranium source) and entering adjacent sedimentary basins, (2) destruction of uranium-bearing monazite, a semi-resistate mineral during deep diagenesis of the sediments of the basin, (3) leaching of uranium from Precambrian basement. Uranium was precipitated (1) by reacting with a graphitic material in the Precambrian basement, and (2) at the unconformity, and (3) above, the unconformity associated with altered zones. These are subclasses of the IAEA deposit type 7. They are here (Table 4.3.1) given deposit type numbers 5.2.1,5.2.2, and 5.2.3, respectively. Fig. 4.3.6B depicts this uranium deposit type, 5.2. The Athabasca Basin of western Canada is the type example of the Unconformity Group. Several other such basins exist worldwide. This is the most important uranium deposit type, as measured by production.

In the Devonian time period, land plants first evolved, and their decay provided organic matter to reduce the hexavalent uranium close to the earth's surface, and in recently buried sediments. This produced the sandstone type 5.3, Fig. 4.3.6C, and other varieties of deposits in sandstones, which are accommodated by initiating subtypes 5.3.1. basal channel deposits, 5.3.2. tabular deposits, 5.3.3. roll-front deposits, 5.3.4. tectonic-lithological deposits, 5.3.5. mafic dykes/sills. Other subtypes could be added. Sandstone types dominate uranium geology in the United States of America, and many places around the world. The ash-leach theory appears reasonable in many instances. Organic plant material is the most important recognized constituent to serve as a chemical reductant to precipitate hexavalent uranium from groundwater. Hydrocarbon gases such as methane were certainly important in the precipitation of uranium in the South Texas Uranium District, and methane may be identified elsewhere as influential. These types of cratonic cover hosted deposits are well known throughout the world. Uranium is being produced in Niger, Kazakhstan, China, and elsewhere.

The Colorado Plateau of the western United States of America is a well-documented, Mesozoic, productive example. Uranium host facies are most often sandstones, Type 5.3, but they change to the west to the limestone collapse breccia pipe hosted type, Type 4.1. The South Texas Uranium District is an example of Deposit type 5.3, on a Cenozoic passive, prograding, continental margin. Ash-flow tuffs have too often been labeled only as "clay" in stratigraphic descriptions and coal beds. The West Texas Volcanic Field was the uranium source for the South Texas Uranium District. It is below suggested that it was also the source of uranium in lignite (north of Houston), and distal phosphate in Florida.

Surface processes

The uranium geochemical cycle continues, entering a special part of the sedimentological phase, the atmosphere and the surface of the earth. Uranium sources presented earlier in intra-cratonic Sedimentary Basins continue here at the surface. There remains, however, one very significant transport mechanism of uranium. The volcanic ashes deposited close to the caldera as ash-flow tuffs are recognized as a uranium source rock, the uranium being in the glass. But those rocks are only a part of the volume of ash erupted from the caldera. One thousand cubic kilometers is a reasonable average eruption volume, and a part of that enters into the upper atmosphere. How much uranium is within 300 cubic kilometers of ash, if a slightly enriched magma might have 10 ppm uranium? Furthermore, calderas occur in clusters, with huge ash production over extended periods. Fig. 4.3.6D suggests the fate of portions of that ash. Ash may remain entrained for some period of time, until certain meteorological conditions are met. Airborne deposition may take place over time and a large area. Meteorological variables may determine preferred directions. It is the ashes from the uraniferous calderas generated on rift shoulders which carry a characteristic chemistry. The uranium deposit types in line to receive uranium (Fig. 4.3.6D) from volcanic ash are calcrete, lignite, (coal), black shale, and phosphate rock (phosphorite). The association of calcretes with arid climates, shallow groundwater, and evapotranspiration suggests they are sourced more by uranium in groundwater. Coal is included here with lignite as an ash recipient, because it forms at the surface in lacustrine and deltaic swamp environments. Lignite can change into coal during diagenesis. During this event the mobile uranium is lost; however the much less mobile REEs also present in the ash or clay layers are not lost. This behavior can be tracked; environments or uranium deposit types subjected to deposition from volcanic ash will have thorium/uranium values that lie along the magmatic trend. The extraction of uranium from sedimentary phosphate rock has been taking place for 40 years. Sedimentary phosphate deposition takes place most commonly near the outer edge of the continental shelf, where cold, deep, nutrient-rich currents are forced up to near the ocean surface where they mix with other water, and the nutrients feed a vast fish population. Fish bones, calcium phosphate, and phosphatic microorganisms are the initial source of the phosphate, but they may undergo several stages of diagenetic changes. They often have a facies relationship with black shales. Black shales are variable; they can be immense thin events or concentrated more locally in thick basins. They represent chemically reducing conditions and strong organic productivity.

Black shales also have sedimentological relationships with other categories in this list of Uranium Deposit Types. They can interface with black sands, coal, lacustrine, and phosphorites. All of these different facies report trace element variability of the same trace elements, uranium being one of them. Large volcanic ash eruptions are suggested to be responsible for variable trace element enrichments. The intra-cratonic basin is the same, except here only the top surface of that sedimentological deposit is considered. Marine phosphate deposits have a facies relationship with marine black shales, and they have high but variable amounts of distinctive trace elements, including uranium, selenium, arsenic, and REEs.

To conclude the discussion of Fig. 4.3.6D, we have included a speculative possibility that ash-fall into the ocean might serve as a transport pathway to mineralize oceanic-platform phosphorites and black shales. Many phosphorites appear to have received their uranium and REEs simultaneous with formation of the phosphate, thus probably under shallow ocean water. But what about the possibility of the ash which fell in the ocean, and currents may have swept the uraniferous plume to shelf phosphates and black shales which would preferentially adsorb the uranium?

This possibility is suggested by the example of a 28–34 MY ash-flow tuff event that consisted of seven major caldera eruptions in West Texas. Downstream and downwind enormous deposits of Catahoula formation of ash and clastic deposition took place, the South Texas Uranium District formed, and meanwhile, ash deposition did not stop at the shoreline. The Gulf of Mexico surface currents go generally clockwise around the Gulf. It is to be tested if uranium dumping into the ocean from the ashes off the Texas coast got swept around to the shallow west Florida shelf where the absorptive properties of phosphorites were operative. Thorium may suggest an answer. What is the uranium/thorium value for Florida phosphate rock; it would suggest airborne ash or water-borne uranium.

Other, outlier, uranium deposit types

Not all uranium deposit types appear to fit into the uranium geochemical flow documented above. Consideration will now be made of these outliers, the last of which is very important. Deposit hosted in carbonatites is an established uranium deposit type, although it has only one good example, Palabora in southern Africa. Carbonatites are better known for thorium resources, and high thorium contents may reflect high uranium. Carbonatites are found in rifts or in long fractures seemingly related to rifts. They can be placed on the left side of Fig. 4.3.2.

6.1 carbonatites. (1.2.3 iaea)
6.2 Convergent-subduction porphyry copper deposit (1.2.1 iaea)
6.3 Convergent- collision
6.3.1 Anatexis (1.1, 5,iaea) 6.3.2 metasomatic (5, iaea)
6.3.2.1 Na metasomatism (5.1, iaea)
6.3.2.2 K metasomatism (5.2, iaea)
6.3.2.3 Skarn (5.3, iaea)
6.3.3 Metamorphic (6 iaea)
6.3.3.1 Strata-bound (6.1, iaea)
6.3.3.2 Structure bound (6.2, iaea)
6.3.3.3 Marble hosted phosphate (6.3, iaea)
6.4 Strike slip –
6.4.1. Regional lateral fault (8, iaea, 2004)
6.4.2. Pull apart (3 iaea)(Olympic Dam)

The strata and structure bound deposit types occur in convergent plate tectonic environments, with porphyry copper deposits forming above subduction zones either in an island arc or at a continental margin, with more uranium favored in the latter (Fig. 4.3.7). Anatexis is favored in continent–continent collisions, along with metasomatic and some metamorphic deposit types. Large strike-slip fault systems are found in cratons, with a single fault or a small family present, often separating different terranes. Perturbations in a family of parallel strike-slip faults often lead to the formation of pull-apart basins, indicating a potential relationship between uranium deposit types 6.4. Cratonic corners are sometimes progressively sheared by large lateral or strike-slip families (Fig. 4.3.7B).

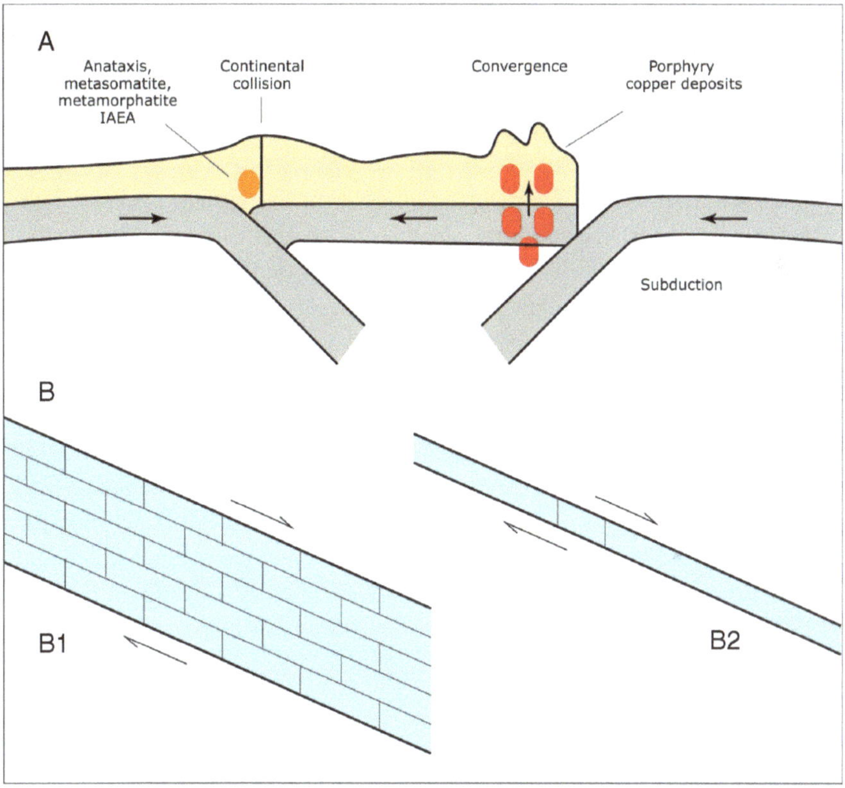

Fig. 4.3.7 ***(A) Plate tectonic cross-section; oceanic plates below, with arrows showing movement.*** Plate tectonically, two interactions are taking place, #1 on the right is plate convergence with subduction of an oceanic plate, producing porphyry copper deposits in the continent of the overriding plate. Uranium participates as a trace element in the copper deposits, but the high volumes of production make this low enrichment being a potential asset and uranium deposit type. On the left-hand side of the (A), two continents are colliding due to plate tectonic convergence. Vast melting, anatexis, metasomatism, and adjoining metamorphism take place. Uranium participates and gets concentrated in albitites (Na metasomatism), and in anatectic, metasomatic, hydrothermal, and metamorphic, events. (B) Map of large shear zones stretching across a craton, the multiple, semi-parallel, faults, which often occur in families. Pull-apart basins can be present, particularly if inflections exist in the strike direction.

In Egypt, miscellaneous pods of uranium enrichment are found along large strike-slip faults. Uranium deposit type 6.4.2 (IAEA class 3) was created to accommodate the Olympic Dam deposit into this body of knowledge. Olympic Dam is in South Australia and is the largest mineral deposit known. Copper and gold are the principal economic objectives; however uranium and REEs are recovered as byproducts, thus it is a significant uranium producer. Its origin is controversial; however it is here thought that the Stuart Shelf (Precambrian) was on a cratonic corner with the Torrance Hinge Line and Adelaide Basin to the east, and a family of large northwest trending, strike-slip faults developing pull-apart basins on the shelf. The Olympic Dam deposit may occupy one of those basins.

Deposits in Zambia, Democratic Republic of the Congo, and the Aillik Group on Labrador, Canada, may also belong to this category. This deposit type is represented and characterized by a single deposit that was found at Roxby Downs/Olympic Dam, in South Australia. The discovery, development, and richness of Olympic Dam have necessitated a completely new uranium mineral deposit type, Iron oxide copper gold ore deposits (IOCG) or Breccia Complex Type. Very few deposits are analogues of Olympic Dam, although many claims to be analogues. This classification is based on multiple parameters. The Oz deposit is a new discovery on the Proterozoic Stuart Shelf near Olympic Dam. Subtype 6.4.2 is here determined to be large pull apart systems illustrated by Olympic Dam. Subtype 6.4.1 is determined to be veins and vein/dyke pairs within regional lateral fault families, as in Egypt.

CHAPTER 4.4

IAEA classification

Table 4.4.1 gives the recent IAEA classification (IAEA, 2009) for comparison. It was not initially constructed for geological purposes, nor was items numbered for better communication. Numbers for IAEA uranium deposit types are only given here. The Geosystems Model is guided by geological ideas, and taking advantage of certain features, such as the surface of the earth, has provided integration and simplification. Classification criteria are both genetic and geometric. The IAEA classification is fickle; it changes with different advocates' what happened to fetid limestones and veins as deposit types? Why are there coal, lignite, and peat? The Geosystems Model does not eliminate any IAEA subclasses, but associates several.

Table 4.4.1 IAEA (2013) (IAEA, 2009) model numbers and titles, correlated with IAEA (2004) and Geosystems classifications.

IAEA (2013)	Geosystem class	Geosystem subclass
1. Intrusive	6	
1.1 Anatectic		6.3
1.2 Plutonic		
1.2.1 Porphyry Cu		6.2
1.2.2 Peralkaline		
1.2.3 Carbonalites		6.1
2. Granite related	6	M1.1, 1.2
2.1 Endogranite		
2.2 Perigramite		
3. IOCG, Olympic Dam	5	6.5
4. Volcanic	9	V2
4.1 Structure bound		
4.2 Strata bound		
4.3 Volcaniclastic bound		
5. Metasomatic	11	M1.3, 6.3
5.1 Na metasomatism		
5.2 K metasomatism		
5.3 Skarn		
6. Metamorphic	12	6.3
6.1 Strata-bound		
6.2 Structure bound		
6.3 Marble-hosted phosphate		
7. Proterozoic unconformity	1	B5.2
7.1 Unconformity contact		
7.2 Breccia or fracture controlled		
7.3 Strataform-structure controlled		
8. Collapse breccia pipe	8	P4.1
9. Sandstone deposits	2	B3
9.1 Basal channel		
9.2 Tabular		
9.3 Roll front		
10. Early Precambrian quartz pebble conglomerate	3	B5.1
10.1 Monomineralic		
10.2 Polymineralic		
11. Surficial deposits	10	
11.1 Peat-bog		B5.4.2
11.2 Fluvial valley		
11.3 Lacustrine - playa		B5.4.1
11.4 Pedagenic and surface fill		
12. Lignite-coal deposits	13	B5.4.2

(Continued)

Table 4.4.1 *(Cont'd)*

IAEA (2013)	Geosystem class	Geosystem subclass
12.1 Startiform syngenetic		
12.2 Mixed strata/fracture epigenetic		
13. Karst in limestones		
13.1 Strata-bound		
13.2 Cataclastic		
13.2 Karst		
14. Phosphate	7	B5.4.4
14.1 Organic		
14.2 Minerochemical		
14.3 Continental		
15. Black shale	14	
15.1. Stratiform		B5.4.3
15.2. Carbonaceous shale/stockwork		P4.3
16. Vein deposits	4	M,V.H.

CHAPTER 4.5

Uranium and thorium geochemistry

Discussion up to now has not been geochemical, by choice. Periodically, in passing, geochemical properties have been mentioned. Now, the use of a simple but powerful concept will illustrate many parameters and properties of uranium deposits. With the *x*-axis being log uranium concentration in ppm and *y*-axis being log thorium concentration in ppm, $y = 4x+4$, expresses the most common magmatic trend, it being outward along this line; this is a generalized magmatic evolution with increasing silica. Magmas are varied. Much more needs to be understood. Some things are clear. Magma and pegmatite chemistry can be high in uranium. More syenitic compositions tend to be higher in thorium. They will generally have both uranium and thorium. Several magmatic regions are shown on the diagram. Felsic volcanic ash is indicated, and deposits that probably get mineralized from felsic volcanic ash, air-fall, will also be close to this line, that is black shale, peat, lignite, coal, and phosphorite. At this point we are attempting to decipher different uranium deposit types by their thorium and uranium properties. Vast amounts of data exist. Simplifications and generalizations are required at this time. Other authors (Mercadier et al., 2011) set out to do this for REE elements. They present data for six deposit types with positive but limited results. Scientists now test the possibility of extracting REEs from fly ash, the solid byproduct of coal-fired electricity plants. Where did those REEs come from? Possibly, from felsic air fall ash.

Results from this initial study suggest (1) important magmatic trend exists, (2) several magmatic digressions exist, (3) as illustrated in Fig. 4.5.1D, volcanic ashes from this magmatic trend may fall on surface geological environments (calcrete, peat, lignite, black shale, phosphorite,) and thus have this geochemical fingerprint. Uranium mobility at some point will tend to destroy part of this information. Peat and lignite are considered Uranium

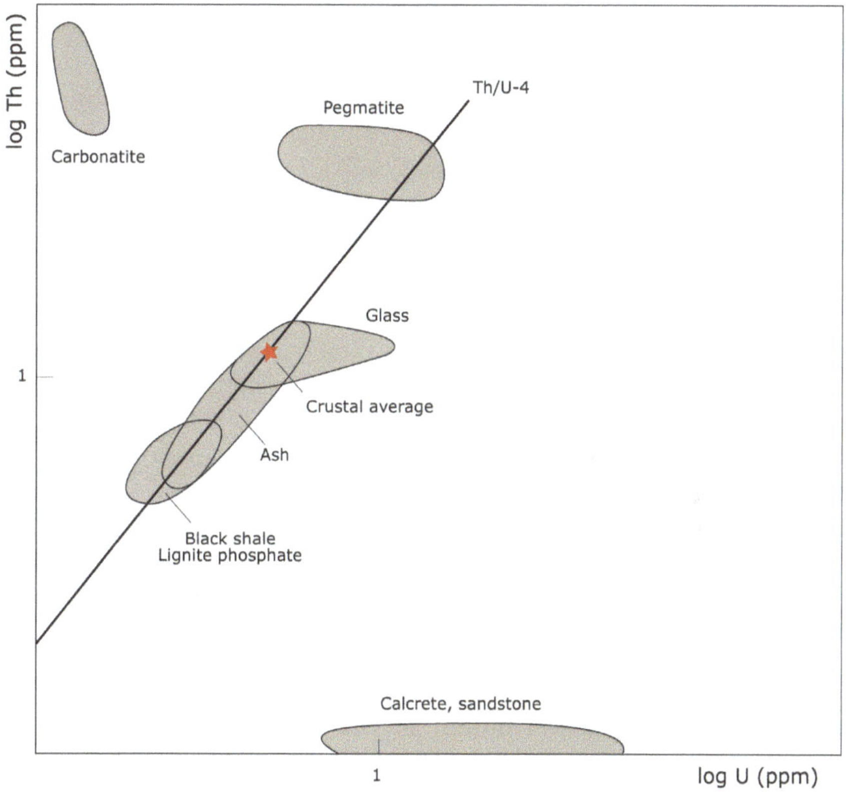

Fig. 4.5.1 ***Uranium and thorium geochemistry plays a significant part in understanding uranium mineral deposit origins.*** Processes start with the *x*-axis being log uranium concentration in ppm and *y*-axis being log thorium concentration in ppm, $y = 4x + 4$, expresses the most common magmatic trend, it being outward along this line; this is generalized magmatic evolution Magmas are varied. Silica-poor igneous rocks lie on the thorium-rich side of this line. Peralkaline rocks have extensions into the uranium-rich side. The ash-leach theory moves environments down to the uranium axis, and to the right, to form uranium deposits. Uranium and thorium get separated in aqueous transport. The ash-eruption transport theory–effected environments will move along the magmatic trend and demonstrate different relationships between uranium and thorium.

resources by the IAEA. Coal is diagenetically altered lignite, and the mobile Uranium was driven out. Diagenesis drives toward the y-axis (uranium mobile, thorium not). (4) The ash-leach theory involves hug volumes of rift shoulder felsic ash moving on the diagram from slightly high in the magmatic trend = initial ash/glass composition, moving rapidly toward the x-axis. Ash leach drives toward the x-axis.

Postmagmatic geochemical behavior of uranium and thorium are that uranium gets oxidized from +4 to +6 which is soluble, but thorium does not have another valence, and remains insoluble. Uranium and thorium get geochemically separated at the earth's surface, except for the small amount of uranium in resistates with thorium, in the black sands (concentrated accessories). Any uranium deposit formed through dissolved uranium in aqueous transport will have essentially no thorium or REEs. These deposits lie close to the x-axis, and include deposits within the intra-cratonic basin, except quartz pebble conglomerates. Calcretes are also more thought of as forming from ground and surface water, rather than air-fall uranium. Now we can measure it. Phosphate deposits, 5.4.4, have a long history with uranium, and a shorter one with REEs. Thorium/uranium values verify an ash source of exotic chemicals in phosphorites. Phosphorites are sometimes mined for their uranium content and are known to be high in REEs. Uranium and thorium geochemistry plays a significant part in understanding uranium mineral deposit origins and providing distinguishing features.

CHAPTER 4.6

Conclusions

Uranium is a very versatile chemical, and geologically, it is found in many environments. More than a dozen different uranium ore deposit types, and more subtypes, have been recognized and described. About the same number of different classifications of uranium deposit types exist. Presented here, for the first time, is a sequence of uranium deposit types integrated according to the geochemical cycle of uranium, beginning with magmatism at continental rift shoulders. The stipulation and recognition of rift shoulders is unique. A-type magmatism, Mo porphyry systems, rare earth, and critical element enrichments, are often present. Weathering of felsic rocks from continental rift shoulders is a significant source of uranium for intra-cratonic basins.

An ultimate end of uranium transport is the intra-cratonic basin, where deposition has varied since the Archean. Stages are represented by deposit types such as conglomerates in the Archean, unconformity type in the Proterozoic, sandstone type in the post-Devonian, and surficial deposit types (illustrated for today, but of course occurred in the past, and would be buried). Surficial deposit types have the added potential advantage of receiving significant amounts of uranium and REEs from long-distance airborne ash associated with caldera-based ash-flow tuff eruptions, from rift shoulders. The integrated sequence of uranium deposit types will provide for scientific understanding, modeling, and better mineral exploration.

The Geosystems Model/Classification of uranium Deposits is an integrated train of thought. The thorium to uranium ratio as a geochemical tool provides understanding and predictive abilities. More detailed studies will result. Think of the isotopic studies, of these elements and associated elements, including REEs, added to the structure outlined by Geosystems uranium Deposit Type Model/Classification.

References

Bowie, S.H.D., 13-17 April, 1970. World uranium deposits, Proc. Uranium Exploration Geology Panel. Vienna, 23–34 (IAEA-PL-391/19), pp. 23–34.

Cuney, M., 2009. The extreme diversity of uranium deposits. Miner. Depos. 44, 30.

Cuney, M., Emetz, A., Mercadier, J., Mykchaylov, V., Shunko, V., Yuslenko, A., 2012. Uranium deposits associated with Na metasomatism from Central Ukraine: a review of some of the major deposits and genetic constraints. Ore Geol. Rev. 44, 82–106.

Cuney, M., Kyser, T.K., 2009. Recent and not-so-recent developments in uranium deposits and implications for exploration. Short Course, 39, Mineralogical Association of Canada, Ottawa, 272.

Dahlkamp, 1993. Uranium Ore Deposits. Springer-Verlag, Berlin, p. 460.

El Kattan, E.M., El Aassy, I.E., 1992. Structurally controlled radioactivity in the Um-Bogma Area, Midwestern- S'inai, Egypt: the geochemistry and uranium favourablity of the postorogenic granites of the Northwestern Arabian Shield. Bull. Fac. Earth Sci. King Abdulaziz Univ. 6, 195–209.

Fayek, M., 2013. Uranium ore deposits: a review. In: Burns, P.C., Sigmon, G. (Eds.), Uranium: Cradle to Grave. Mineralogical Association of Canada, Canada, 43, pp. 121–147.

Goodell, P.C., 1985. Classification and model of uranium deposits in volcanic environments: In Uranium deposits in volcanic rocks, Proc. Technical Committee meeting. El Paso, 1984 (IAEA; STI/PUB/690), pp. 1–16.

IAEA, 2004. https://www-pub.iaea.org/MTCD/Publications/PDF/Pub1259_web.pdf.

IAEA, 2009. World Distribution of Uranium Deposits (UDEPO) with Uranium Deposit Classification. IAEA-TECDOC-1629, IAEA, Vienna, p. 109.

IAEA, 2013. https://inis.iaea.org/collection/NCLCollectionStore/_Public/45/089/45089795.pdf.

Mercadier, J., Cuney, M., Lach, P., Boiron, M., Bonhoure, J., Richard, A., Leisen, M., Kister, P., 2011. Origin of uranium deposits revealed by their rare earth element signature. Tierra Nova 23, 264–269.

Plant, J.A., Simpson, P.R., Smith, B., Windley, B.F., 1999. Uranium ore deposits-products of the radioactive earth. In: Burns, P.C., Finch, R., Eds., Uranium: Mineralogy, Geochemistry and the Environment. Reviews in Mineralogy, Vol. 38, Mineralogical Society of America, 255–319.

Skirrow, R.G., Jaireth, S., Huston, D.L., Bastrakov, E.N., Schofield, A., van der Wielen, S.E., Barnicoat, A.C., 2009. Uranium mineral systems; processes, exploration criteria and a new deposit framework. Rec. Geosci. Aus. 44.

US Geological Survey, 2010, Mineral deposit models for resource assessment: Scientific Investigations Report 2010-5070. https://pubs.usgs.gov/sir/2010/5070/a/.

Wenrich, K.J., 1985. Mineralization of breccia pipes in northern Arizona. Econ. Geol. 80 (6), 1722–1735.

West, M., Ni, J., Baldridge, W.S., Wilson, D., Aster, R., Gao, W., Grand, S., 2004. Crust and upper mantle shear wave structure of the southwest United States: Implications for rifting and support for high elevation. J. Geophys. Res. 109, B03309.

CHAPTER 5

Uranium resources in the Middle East and North Africa

Introduction

The regions of the Middle East and North African countries occupy an important area from the strategic position, where they form a part of West Asia Continents and North Africa, respectively (Fig. 5.5.1), and are situated on the Arabian and Persian Gulfs and the Mediterranean and Red Seas, respectively. The Middle East countries include Egypt, Iran, Iraq, Jordan, Saudi Arabia, Sudan, Syria, Turkey, United Arab Emirates, Oman, Qatar, Kuwait, and Yemen. The North African countries include Egypt, Algeria, Libya, Morocco, and Tunisia.

CHAPTER 5.1

Recognition criteria for uranium resources provinces

The main criteria needed for mapping uranium provinces rely on the attribute of the host rocks and the nearby geological settings (Salman, 1995). It is vital to emphasize the recognition or identifying criteria for uranium in various lithologies such as granites and sandstones. These features will be used for prospecting potential uranium deposits within the Middle East and North Africa region. These two rock groups are significant for hosting the major uranium deposits in the region (Fig. 5.1.1). The important criteria for characterizing uranium deposits in granitic rocks mainly include uranium concentration, lithology, and formation ages. The average concentration of uranium in different types of granitic rocks should be precisely calculated. The distinction between the enriched uranium and ordinary granites is significant. In Brazil, for example, the mean concentration of uranium in poorly enriched granites is about 3–7 ppm. Exploration geochemical studies showed that 12 ppm is the limit between normal granites and uranium-enriched granites (Tassinari and Barreto, 1992). It was also observed that the analyses of uranium-enriched granitic pluton from the Pan-African Shield has uranium contents between

Uranium Geology of the Middle East and North Africa
DOI: https://doi.org/10.1016/B978-0-323-90992-1.00005-1

Fig. 5.1.1 *Map showing the MENA countries.*

5 and 10 ppm and thorium contents between 15 and 25 ppm, thus giving thorium/uranium ratios of 2–2.5 (Rogers, 1978). The low ratios of thorium/uranium in granites are important due to the fact that most highly segregated granites have a ratio of thorium/uranium of more than 4. These ratios which can be described as low could either come from thorium content decrease or losses from the granites, which is geochemically difficult and unlikely, or possibly uranium addition to the plutons. However, such small ratios also make it unlikely or difficult that uranium in the host or wall rocks surrounding the mineralized plutons was possibly formed from hydrothermal fluids resulted or came from the plutons, a process that should have pushed up the thorium/uranium ratios in the remaining source granite to higher values. Tassinari and Barreto (1992) implied that there is a relationship between the uranium-enriched granitoid and the age of the granitoid, because certain uranium-enriched granitoids are limited to a certain age group. For instance, the time period between 1800 and 1300 Ma has been found to be strongly connected with uranium-enriched granitoid. Nevertheless, the time intervals of 2600–2000 Ma, 2200–1800 Ma, and 900–500 Ma are other periods where uranium-enriched granites are found. It has become more evident that the various amounts of uranium concentration are related to different granitic lithological groups. The granitic lithologies that are alkaline granites have higher uranium concentrations. In general, the uranium-enriched alkaline granites are mainly made of quartz-syenites, syenites, and are constituted by biotite-hornblende granites. Granite-related uranium-rich veins have been found to form in or outside late magmatically or metamorphically altered peraluminous leucogranites (Dahlkamp, 1987). In the determination of the original sources of uranium-enriched granitoid, the analysis of the initial Sr ratios is significant.

The recognition and identification metrics for uranium provinces in sandstones differ according to the kind of deposit (Dahlkamp, 1987). Uranium presence in continental fluvial arkosic sandstone is mainly almost flat-lying (5°) strata unless post-ore tilted and often interbedded with argillaceous horizons. It often occurs with tuffaceous sediments. A distinction is drawn between (1) Phanerozoic (Post Devonian) plant-related deposits of organic terrestrial origin and (2) Proterozoic deposits of volcanic algae (Rogers, 1978). Two classes of sandstone-type uranium deposits that are called the roll-front deposits are found: continental basins and coastal plains. In the continental basin uranium deposits, the uranium occurs as disseminated deposits at redox boundary in subarkosic sandstones and arkosic and is deposited in inter-mountain or intra-cratonic basins. They are usually proximal to rocks having anomalous uranium content such as granite or tuffs. Several deposits exist in interbedded sequences of volcanic rich sediments and fluvial sandstones without major time or erosional breaks.

CHAPTER 5.2

Middle East countries

The uranium potential and proven resources in the Middle East and North African countries will be discussed next. The countries were arranged in the ascending order, the Middle East countries include Egypt, Iran, Iraq, Jordan, Saudi, Sudan, Syria, Turkey, United Arab Emirates, Oman, Qatar, and Kuwait as well as Yemen. The treated North African countries include Algeria, Libya, Morocco, and Tunisia. The geological outlines, a historical review of uranium resources, uranium prospects, and uranium potentialities in each country will be discussed.

CHAPTER 5.3

Egypt

Geological overview

Egypt contains various lithological types that range from Precambrian basement lithologies and a thick Paleozoic to recent sedimentary cover. The Precambrian basement rocks cover about 100,000 sq. km in area and can be found in the south Sinai, Eastern-Desert, and with limited outcrops in the south Western Desert Fig. 5.3.1. The basement rocks are grouped into Pre-Pan-African rocks comprising higher grade metamorphic rocks, Pan-African rock assemblage comprising amphibolites and the related island arc association, and Phanerozoic alkaline rocks (EI Gaby et al., 1990).

The thick sedimentary sequence includes a variety of sandstones, limestones, and shales deposited in different geological episodes ranging from the Paleozoic to recent. The Paleozoic sequences are mostly continental clastic interbedded with marine lithologies and then grading upward to carbonate and fine clastic lithologies. The Mesozoic and Cenozoic sequences are mainly clastic with some evaporites and minor marine interbedded sequences. The classification and the main characteristics of the Precambrian-basement rocks in Egypt are shown in Table 5.3.1. Uranium exploration started in Egypt in 1956. The exploration included integrated airborne and ground radiometric prospecting. These methods discovered several radiometric anomalies and uranium occurrences in various

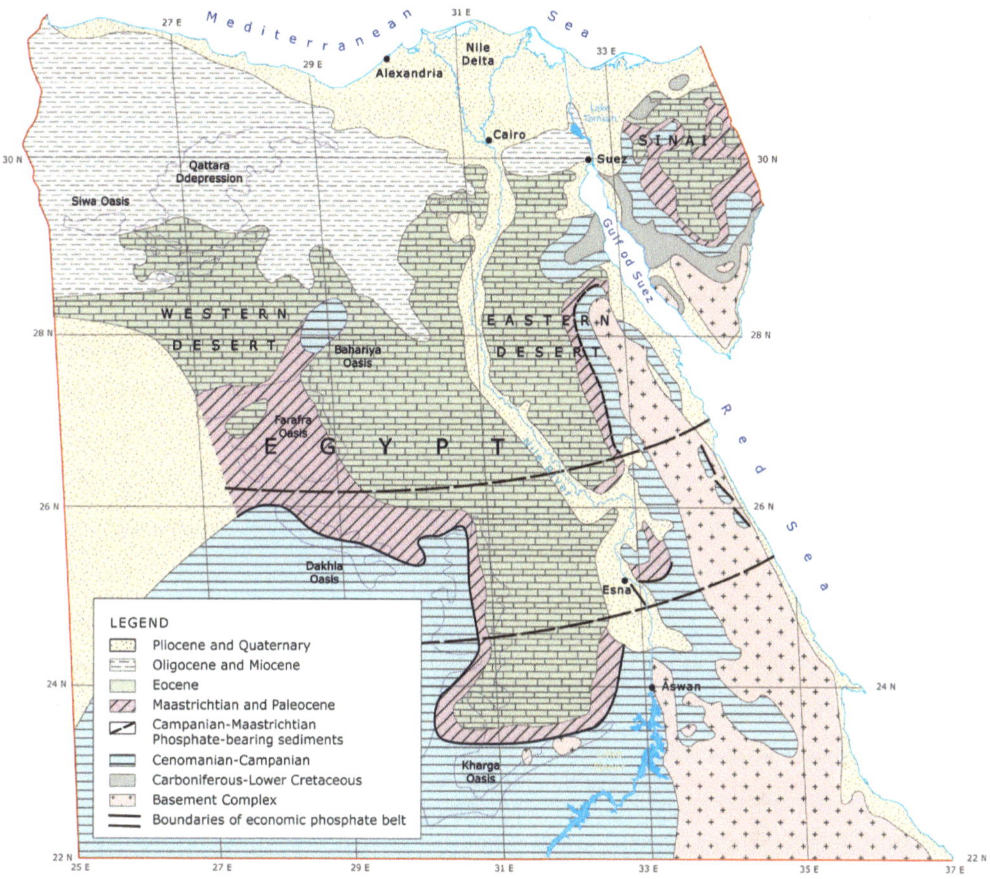

Fig. 5.3.1 ***Simplified Precambrian basement exposures map of Egypt and north Sudan (After El Gabby et al., 1990 with modifications).***

geological environments. The most promising uranium occurrences were found in the younger postorogenic granites and Paleozoic clastic sediments.

Historical review

Egypt commenced active exploration program for Uranium since the early sixties. The major investigating methods at that time were airborne, car-borne surveys over sub-crop or outcrop terrains. Dozens of radioactive anomalies have been detected in different geological environments. These regional investigation endeavors yielded several uranium occurrences in granites of the Late Proterozoic age. Other occurrences have

Table 5.3.1 Classification and characterization of the shield rocks of Egypt showing the uranium bearing units.

No	Rock units and localization	Uranium resources
VII	Intraplate magmatism and sedimentation:	
	Mafic, intermediate and felsic dykes; basalt, dolerite, andesite and rhyolite (Allover Egypt in Shield domains)	Uranium and uranium/thorium
	G3 granites, Alkalic-peralkalic (Eastern Desert (ED) and Sinai), (ED and Sinai).	Uranium
	G2 granites, Calc-alkalic, biotite granites monzogranites, syenogranites, and leucogranites (Allover Egypt in Shield domains). Ultramafic-mafic layered intrusions dunite, peridotite, pyroxenite, gabbro, norite, troctolite and leucogabbro (Younger Gabbros); calc-alkalic and alkali (ED and Sinai).	Uranium and uranium/thorium
	Clastic molasse sediments (Hammamat Sediments) poorly sorted polymictic conglomerate, immature greywacke and limestone with minor calcareous beds; deposited in intermontane basins. Rhyolite (post–Hammamat felsite), (ED and Sinai).	
		Uranium
VI	Continental margin volcanics: basaltic andesite, andesite, rhyodacite and rhyolite. Calc-alkali – alkali; includes dokhan volcanics and some felsites (rhyolites), (ED and Sinai)	Uranium
		Uranium and uranium/thorium
V	Subduction–related granitoids (arc granitoids):	
IV	Arc meta-volcanics: intermediate-felsic metavolcanics and metavolcanic-clastics. Meta-andesite, metadacite, metarhyo-lite, metatuffs and occasional Banded Iron Formations (BID) (Central Eastern Desert (CED) and Southern Eastern Desert (SED))	
III	Ophiolitic Melange: ophiolitic fragments of unit II in a matrix of paraschists of pelitic and psammopelitic composition. Mostly green schist facies and rarely amphibolite facies metamorphism, (CED and SED)	
		Uranium/thorium and uranium
II	Ophiolites:	
I	Gneisses and related rocks:	

been mapped in clastic sediments of Paleozoic age. The Nuclear Materials Authority (NMA) represents the governmental body in charge of uranium exploration in Egypt and has concentrated its main exploration activities in known mineralized uranium locations in the Sinai and Eastern Desert (Uranium, 1999). These exploration activities include detailed geological surveys, structural and radiometric mapping, deep trenching, core and percussion drilling, exploratory tunneling. The activities also include sampling of the uriniferous lenses for laboratory analyses and technological studies for defining the characteristics of the mineralization. It is known that Egypt paid a great attention since early times to the placer of black sand deposited in the Sinai coast and North River Nile Delta. These deposits are present in the form of coastal black sand sheets and an ENE-WSW trending sand dune belt. The mineralogical analyses of these sand deposits indicate the presence of zircon, monazite, and rutile in addition to magnetite and ilmenite. Based on NMA efforts since the 1960s, a number of uranium prospects were discovered. The characteristic features of these uranium prospects and their geological environments will be discussed next.

Uranium prospects

El Missikat and El Erediya

Many of the younger granites in the Egyptian region are of the uraniferous type, where normal uranium content ranges from 10 to 35 ppm (Bakhit and Guirguis, 1983; Salman et al., 1989; Abdel Monem et al., 1990). The El-Erediya pluton was formed between 603 and 575 Ma. It is located near the northern central block edge of the Eastern Desert which has oceanic exposed rock (Greenberg, 1981, Stern, 1981; Stern et al., 1984). The El-Erediya and El-Missikat uranium deposits are among the oldest uranium occurrences in Egypt and include primary and secondary uranium minerals (Attawiya, 1983). The host rock is mainly granites that are exposed in the Egyptian central Eastern Desert. The outcropping granite is seen by topographic relief with steep slopes (Figs. 5.3.2 and 5.3.3). This granite was formed during the Pan-African orogeny and is associated with older granites forming the main country rocks in this region (Fig. 5.3.4). The uranium mineralization at El Missikat is hosted in ENE trending shear zones (Fig. 5.3.5) within orthoclase granite. It is connected with jasperoid silica (Abou Deif, 1985) and there is a strong alteration represented by

Fig. 5.3.2 ***Central Eastern Desert showing El Missikat (1) and El Erediya (2) granite exposure, Egypt.***

silicification, sericitization, hematitization, and kaolinization. The uranium minerals are mostly soddyite and uranophane with fine disseminated sooty pitchblende. These minerals are associated with sulfide and gangue minerals that include chalcopyrite, pyrite, sphalerite, galena, and molybdenite. The gangue minerals are mainly manganese oxides, iron ore, and fluorite (Hussei and Sayyah, 1992). Uranophane and beta-uranophane are the most common secondary uranium mineral.

Fig. 5.3.3 ***General view for El Missikat prospect.***

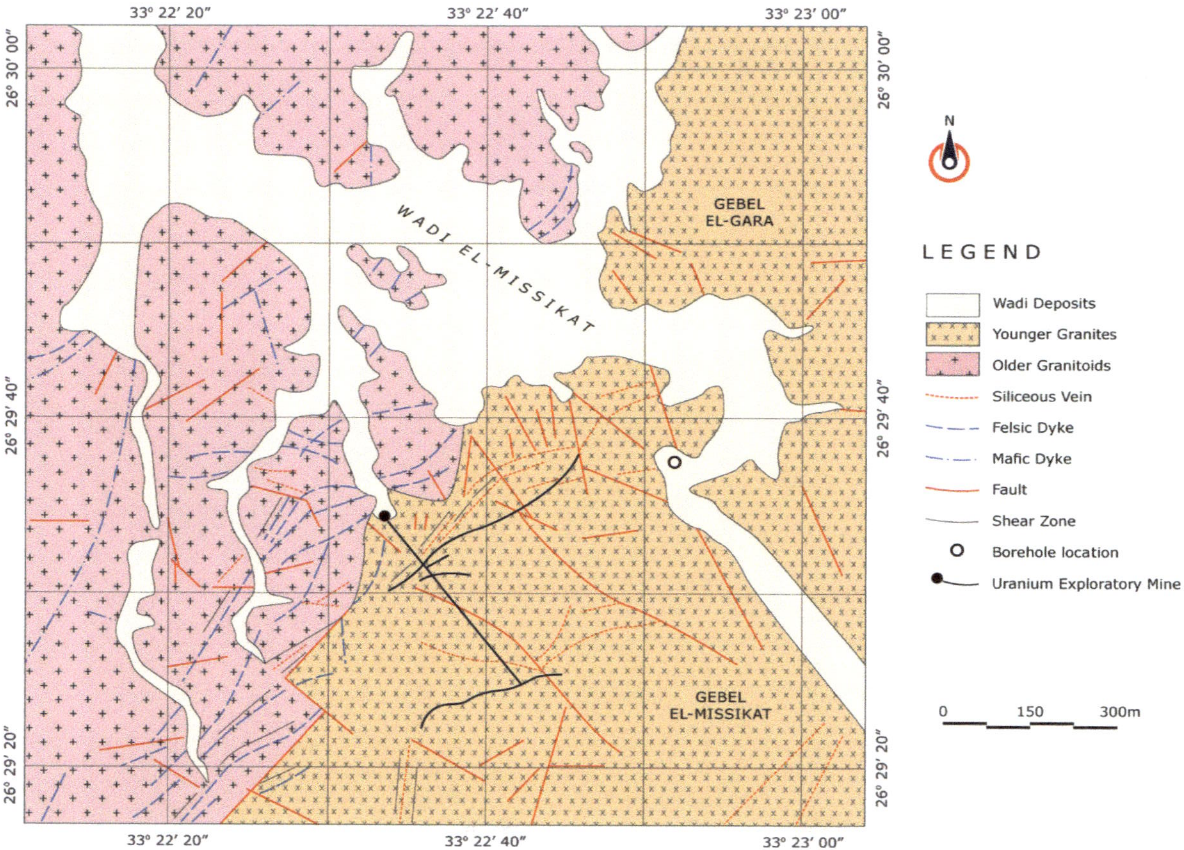

Fig. 5.3.4 ***Geological map for El Missikat locality showing the main rock units and some structural features which mainly follow ENE-WSW and NW-SE trending shear zones and faults, Central Eastern Desert, Egypt.***

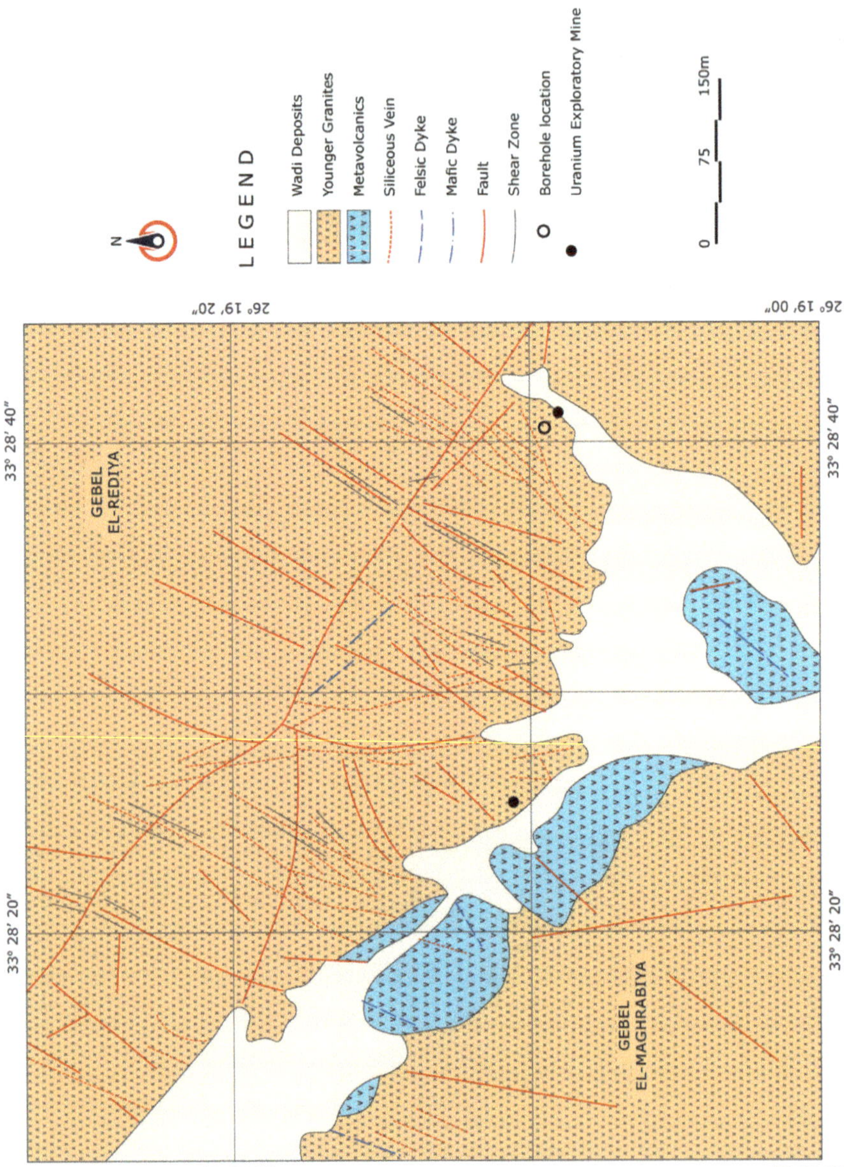

***Fig. 5.3.5** Geological map for El Erediya locality showing the main rock units and some structural features which mainly follow ENE-WSW and NW-SE trending shear zones and faults, Central Eastern Desert, Egypt.*

Um Ara

The Um Ara prospect is in the southern Eastern Desert, southeast of Aswan City. Two types of granites are exposed in Um Ara locality: monzogranites and microgranites, and the transitional contact between the two types of granites was derived first from fractional crystallization and then by metasomatic hydrothermal alterations. Both granites are epizonal, non-foliated, and belong to anorogenic A-type Egyptian younger alkali granites. Mylonitization, metasomatic alterations, and high radioactivity characterize the microgranites compared to the parent monzogranites (Osman et al., 2001). Radioactive mineralization in the younger Um Ara is restricted to the micro- to medium-grained varieties that range in composition from albitized to the potash feldspar-rich granites. Uraninite and/or thorite inclusions in orthoclase, plagioclase feldspars, quartz, and biotite aid in determining their syngenetic origin. A later alteration caused the oxidation to secondary minerals including uranophane and curite (Abdel Meguid, 1986). These minerals are associated with violet fluorite. The wall rock alteration is composed of silicification, microclinization, albitization, and hematitization. The uranium mineralization of Urn Ara is structurally controlled following WNW and NW trends (Hussein et al., 1986).

The granite plutons are thought to have a high potential for workable uranium deposits that belong to a vein type. Despite the high uranium values in these postorogenic Pan-African granites, there have been no large concentrations of magmatic uranium found to date. Moreover, Struckless et al. (1983) found that the low thorium/uranium and high uranium/potassium ratios argue against any concentrating of uranium by magmatic processes in the post orogenic granitic rocks of the northwestern areas of the Arabian Shield. The hydrothermal vein–type uranium deposits of low-to-medium scale have been identified in the marginal zones of these granite plutons. These deposits were found based on exploration and the presence of various types of uranium occurrences (Salman et al., 1990). The persistence of the secondary uranium mineralization for more than 300 m depth, with the increase in the width of the shear zones with depth, existence of strong hematitization, partial silicification and fluoritization, and the presence of pitchblende in some underground works provide evidence of potential uranium deposits.

Jebel Qattar prospect

Jebel Qattar is located at the northern Red Sea Hills and it is one of the economically more promising younger leucogranite exposures in this area

Fig. 5.3.6 ***Space image for west Hurghada area showing the main exposures, North Eastern Desert.***

(Fig. 5.3.6). It is characterized by a rugged topography with steep slopes (Fig. 5.3.7). This nature of this topography makes uranium exploration difficult. This granite is calc-alkaline and was produced by partial melting of the upper mantle or lower crust (Greenberg, 1981; El Shatoury et al., 1984). The granite is coarse-to-medium grained, varying in color from pink to reddish pink in fresh samples, but it tends to turn pale or reddish brown in color especially along the shear zones and intersections of fractures. The most characteristic alteration features are hematitization, silicification, kaolinization, fluoritization, and epidotization. The uranium mineralization is mainly hosted in the younger biotite granite. This granite is essentially composed of quartz, orthoclase with subordinate oligoclase, and biotite. The accessory minerals are represented by zircon, hematite, sphene, and apatite. The uranium mineralization is found within fault zones and smaller scale fractures that follow NNE, ENE, and NW trends, where the mineralization is present in patchy or within lens (Salman et al., 1990). The associated alteration features (Fig. 5.3.8) with uranium mineralization represented by intense hematitization, partial kaolinization, light carbonatization, and violet fluorite are frequently present. The identified secondary uranium minerals are: uranophane, carnotite, clarkeite, kasolite, zippeite, soddyite, and primary uraninite (Sayyah and Attawiya, 1989). The secondary uranium mineralization is found as a staining film along the fracture's surfaces in the mineralized shear zones that follow NNE–SSW in the granite pluton of

Fig. 5.3.7 ***Jebel Qattar, showing rough topography, looking north.*** *Q1 and Q VI are mineralized shear zones.*

Fig. 5.3.8 ***Uranophane-forming thin films along fracture surfaces, associated with strong alterations, Jebel Qattar prospect, Eastern Desert, Egypt.***

Fig. 5.3.9 ***Poly-metallic altered mylonites, Abu Rusheid locality, south Eastern Desert, Egypt.***

Jebel Qattar. The uranophane and beta-uranophane are the most common secondary uranium minerals (Fig. 5.3.9).

Uranium resources potentialities

Proterozoic basement rocks

When exploration for uranium was started in Egypt, there were no records of uranium prospects in the oldest Proterozoic basement rocks, except radiometric anomalies within the Abu Rusheid area in the southeastern Desert of Egypt. These anomalies were found during a 1968 airborne radiometric survey performed by the Lockwoods Survey Corporation. During the geological groundwork of this anomaly, evidence of metasomatic processes was found accompanied with tantalum-niobium mineralization (Hassan, 1973). This rare metal mineralization is believed to be associated with sill-like body of albitized and amazonite granite and Sabet et al. (1976) considered it to be formed during the successive stages of metasomatic alteration of quartz-feldspar gneiss and granite gneiss. Several studies have considered this gneiss to be a metamorphosed-sedimentary unit of quartzo-feldspathic composition (Hassan, 1964; Abdel Monem and Hurley, 1979; El Ramly et al., 1985; and Al-Saleh, 1998). Ibrahim et al. (2001) considered that Abu Rusheid gneiss is a highly mylonitic gneissic granites, ranging in composition from granodiorites to adamellites, which were subjected to intense pressure at its contact (thrust fault) with the ophiolitic mélange rocks in the Abu Rusheid–Sekeit area (Figs. 5.3.10 and 5.3.11).

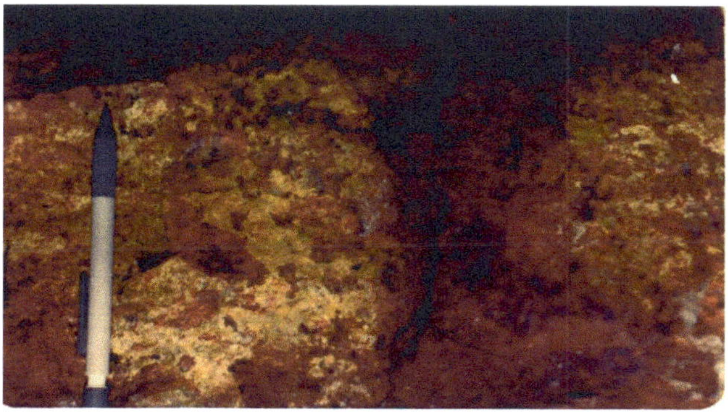

Fig. 5.3.10 ***Close view of the poly-metallic altered mylonites, Abu Rusheid locality, Close-up view, south Eastern Desert, Egypt.***

The cataclastic rocks in this region are characterized by high radioactivity readings and have been considered as mineralization hosting geological units. These rocks were dislocated by a NNW-SSE trending shear zone which is highly hematitized Fig. 5.3.11. The wall-rock alteration is commonly potassic and sericitic with the potassic alteration consisting of magnetite, hematite, allonite, k-feldspar, and albite. However, the sericitic alteration includes hematite, sericite, carbonate, and fluorite. Four types of sulfides are common: fine-grained pyrite bornite, arsenopyrite, chalcopyrite, and sphalerite. Recently, new uranium discoveries have been found within quartzite rocks at Wadi Sekeit in the south Eastern Desert.

Fig. 5.3.11 ***Highly radioactive main shear zone, Abu Rusheid locality, south Eastern Desert, Egypt.***

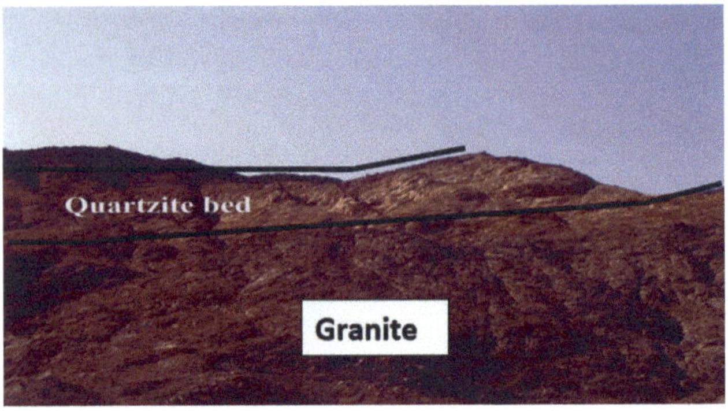

Fig. 5.3.12 ***Uranium-bearing quartzite, Wadi Sakait, south Eastern Desert, Egypt.***

These quartzites extend 2 km with an approximately 0.6 km areal extent (Fig. 5.3.12). The quartz is whitish color, hard and compact with some iron patches, and occasional presence of yellowish secondary uranium minerals. This quartzite bed is often a roof pedant to agranitic outcrop, which represents the country rock. The preliminary chemical results indicate that this uranium mineralization is of low grade, where the uranium concentration ranges from 200 to 600 ppm. The Proterozoic basement rocks especially the quartzite and altered mylonites have a large uranium resources potential in Egypt. These rocks are mainly exposed in the South Eastern Desert. This region should be explored in detail for uranium, base metals, and rare earth elements. It has a high potential to find an economic uranium deposits in such geological environments. Moreover, it also has potential to find some uranium deposits in some unconformity zones within these old basement rocks.

Younger granites

The younger granites cover a large portion of the Egyptian deserts. There have been numerous radioactive anomalies found in these younger granites. Some of these anomalies are now uranium prospects that can often represent potential areas for economic uranium resources. The uranium mineralization is mainly associated with shear zones, fractures, and joints. These shear zones are mainly found in the trending NNE-SSW and ENE-WSW directions in many locales including Jebel Qattar, and El Erediya and El Missikat areas in the Egypt's Eastern Desert. The uranium mineralization includes secondary and primary types. There is lack of information on the oxidation–reduction boundaries, level of erosion in the granite plutons, and geochemical characteristic changes at deeper levels. The cause of this lack of information is the rough topography and most of the mineralized shear zones are dipping against the slope of the mountains. This makes it difficult to perform drilling programs at reasonable costs. Most of these occurrences need an extensive underground working for the exploration and exploration drilling, and to study the characteristics of the uranium resources and evaluate the reserves in these areas. An age determination for pitchblende lenses from the El Erediya's younger granite uranium prospect from isochrons age (162 Ma), using $^{207}Pb/^{204}Pb$-$^{206}Pb/^{204}Pb$ and $^{207}Pb/^{206}Pb$-$^{204}Pb/^{206}Pb$ data, is conducted to represent the formation age of the pitchblende deposit at El Erediya (Abu Deif, A., 1992). Evidence for recrystallization and structural deformation of the pitchblende lenses at El Erediya is seen by the fracturing of the pitchblende grains and invasion of fractures by the third generation of silica which carried disseminated pitchblende minute grains. This deformation event took place around 55–45 Ma as evidenced by the age obtained by the uranium-lead isochrons (Abu Deif, A., 1992).

One of the most promising uranium occurrences is located within a major shear zone within two mica monzogranites known as the El Sella shear zone. This shear zone is situated in the south Eastern desert in Jebel El Sella and it is divided to three parts by two wrench faults following N15° W into three separated parts. This zone includes red and gray jasperoid mineralized silica veins with visible pyrite and secondary uranium minerals. Argillization, fluoridization, hematitization, sicification, carbonatization, and sulfidization are the main alteration processes (Ibrahim et al., 2003). Ibrahim et al., showed that the uranium migrated out from the surrounding granite plutons toward the shear zone which represents a depositional site for the migrated uranium. Based on the available information, the following features can represent good targets to find uranium resources in the younger granites (Pan-African):

1. The northern contact zones of the granite plutons with the country rocks especially the metamorphosed sediments and metavolcanics. Many uranium prospects in Egypt are located within the northern zones of these granites.
2. The ENE-WSW and NNE-SSW shear zones and their intersections in these granites represent excellent target for uranium exploration and localization. Moreover, the extension of these shear zones into the adjacent Paleozoic rocks often has good potentials for uranium deposits.

Fig. 5.3.13 ***Granite-basic dyke contact, perfect trap for uranium mineralization, Jebel Qattar, Eastern Desert Egypt.***

3. The areas where basic dykes are located within these granites represent good locations for uranium mineralization (Fig. 5.3.13). These basic dykes act as barriers where uranium-bearing solutions cannot penetrate and uranium minerals are deposited if the geochemical environment is favorable. This scenario occurs at the Jebel Qattar underground exploratory tunnel where secondary uranium mineralizations were densely present at the western contact of the basic dyke with granite.
4. The episyenitic granites found along the contact zones of the granites represent an excellent target for uranium exploration. Many examples of these episyenitic granites are found within Jebel Qattar, El Missikat, and Kabb Ameri areas in the Eastern Desert of Egypt.

Volcanic rocks

There have been several volcanic episodes in Egypt since the Precambrian, especially in the Mesozoic in southern Egypt. The Mesozoic-aged magmatic events can be related to two phases of igneous activity between the Late Jurassic and Early Cretaceous (140 + 15 Ma) and during the Late Cretaceous–Early Tertiary (90+ Ma). The Wadi Natash volcanics represent major magmatic events in the southern Eastern-Desert that belongs to the second phase (Late Cretaceous–Early Tertiary) of the igneous activity (Hashad et al., 1982). The south Wadi El Kareem Hammamat Basin includes numerous bostonite dykes and sills, where the majority of them have radioactivity anomalies (Fig. 5.3.13). The most famous occurrence for

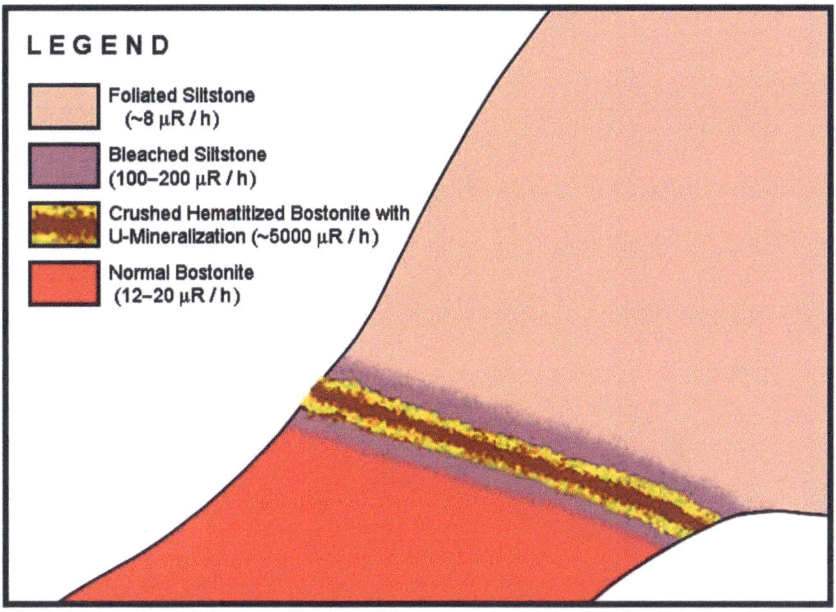

Fig. 5.3.14 ***Schematic cross-section for bostonite-siltstone contact, El Atshan uranium prospect, Central Eastern Desert, Egypt.***

uranium-bearing bostonites is the Wadi El Atshan prospect (Fig. 5.3.14). It includes highly radioactive bostonites with secondary uranium mineralization and pitchblende. The mineralizations are associated with highly altered contact zone within the metasediments (Fig. 5.3.15). The north margin is partly dislocated by an ENE-trending thrust fault (Salman, 1983). The bostonites are found in three generations: two events include radioactive anomalies and uranium mineralization (Ibrahim Salman, 1968 and Salman, 1983). This province is nearly surrounded by numerous post orogenic uraniferous younger granite plutons. This environment indicates that the source rock for uranium is present in addition to the suitable lithostratigraphic units and structural features that can act as hosts for uranium deposits. The volcanic rocks at Wadi Natash, South Eastern Desert of Egypt are interbedded within Nubian sandstones and bear uranium. There is a linear relationship between the volcanic bodies at Wadi Natash and the main structural trends. This area has been affected by three deformational events: (1) NW-trending Precambrian Najd fault system, (2) Upper Cretaceous rejuvenation, doming and transform faults, and (3) Tertiary normal faulting. The basic volcanics within the Wadi Natash area consists of alkaline rocks that were erupted along the rejuvenated transform faults during the second event (Noweir et al., 2003).

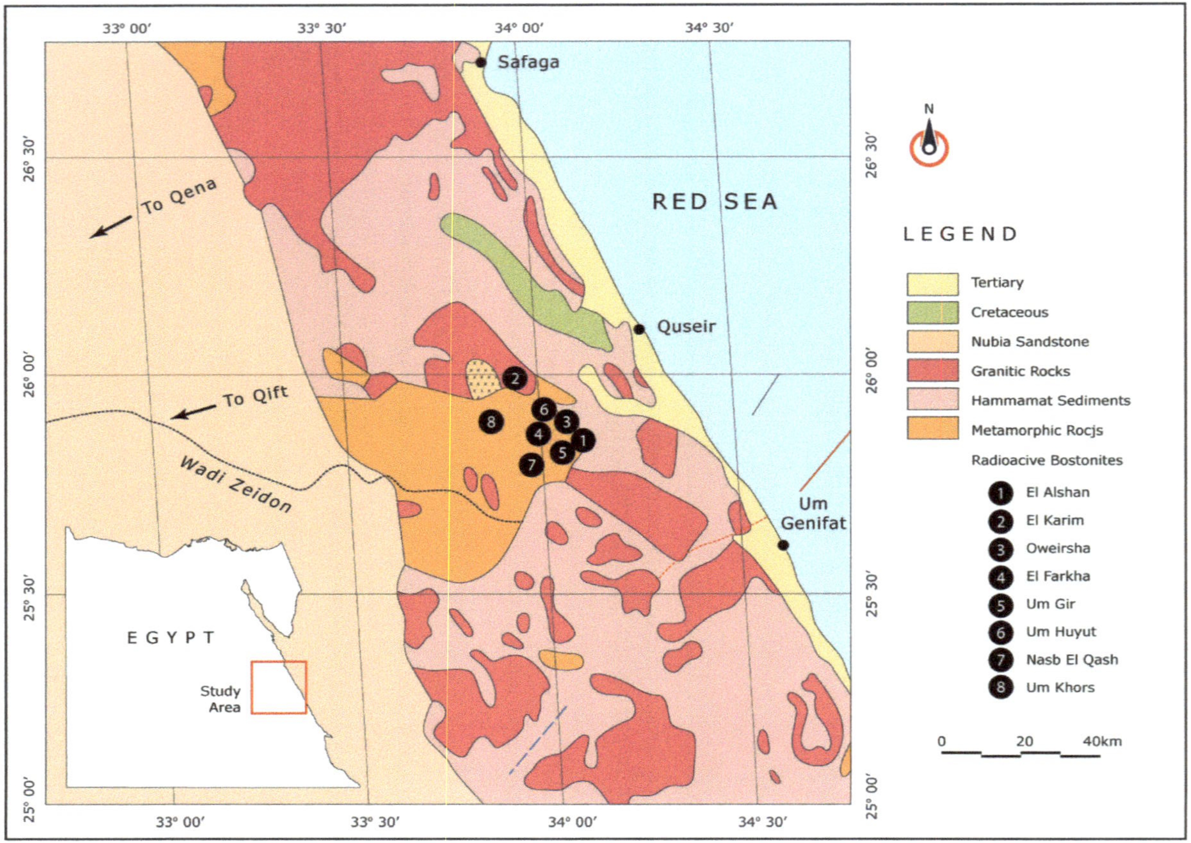

Fig. 5.3.15 ***Distribution of Bostonite dykes in Wadi El Kareem intra-cratonic Hammamat Basin, Central Eastern Desert, Egypt.***

The volcanic rocks within Wadi Natash consist mainly of alkali olivine basalt, hawaiite, mugearite, and trachyte with dolerite being less common. Pyroclastic volcanic rocks locally outcrop. Major element variations indicate that fractional crystallization transformed alkali olivine basalt to mugearite; however trachyte plugs, and dykes might have been originated from the primary alkaline magma similar to that Kenyan portion of the East African rift system (Noweir et al., 2003). It is clear from the characteristic features and distribution of these volcanic rocks that they can form additional resources for uranium. Numerous radioactive anomalies and uranium occurrences have been reported at Wadi El Kareem, Wadi El Owerrsha, El Farkha, Nasb El Qash, and Um Gir (Fig. 5.3.15). The identified secondary uranium minerals are Uranophane, carnotite, clarkeite, kasolite, zippeite, soddyite, and primary uraninite (Sayyah and Attawiya, 1989).

Intra-cratonic basins

From a uranium potentiality point of view, the two Hammamat provinces in the Eastern Desert could be considered as exploration targets. They are located south of Jebel Um Tawat and south of Wadi El Kareem. However, the South Jebel Um Tawat prospect represents an intra-cratonic basin (Fig. 5.3.16), which includes a thick succession of the Hammamat sediments of Late Proterozoic age. The basin is surrounded by Gebel Abu Harba and the Gebel Gattar younger granite plutons on the north side and south side, respectively. These molasse-type poorly sorted clastic sediments are intercalated with minor impure calcareous layers (EI Gaby et al., 1990).

The formations are best developed in the northern part of the central Eastern-Desert and attain substantial development in the north Eastern-Desert and Sinai. The structural features especially the ENE, N-S, NNE, and NW trending faults have a significant role in the formation of the South Jebel Um Tawat basin (Fig. 5.3.17). The uranium mineralization was recorded in the southern zone of the South Jebel Um Tawat basin. These mineralizations are associated with Hammamat sediments, episyenite, and granite. The tectonic and lithological features play the main role in controlling the distribution of uranium mineralization and mode of entrapment.

Two bulk samples have been taken, one from the uraniferous Hammamat sediments and one from the episyenite. Twelve samples have also been collected from rock exposures. Most of the separated fractions of the samples were analyzed for uranium and/or thorium using both radiometric and wet chemical analyses techniques. Also, the oxidation–reduction potential for different rock types was determined. A close relation has been proved

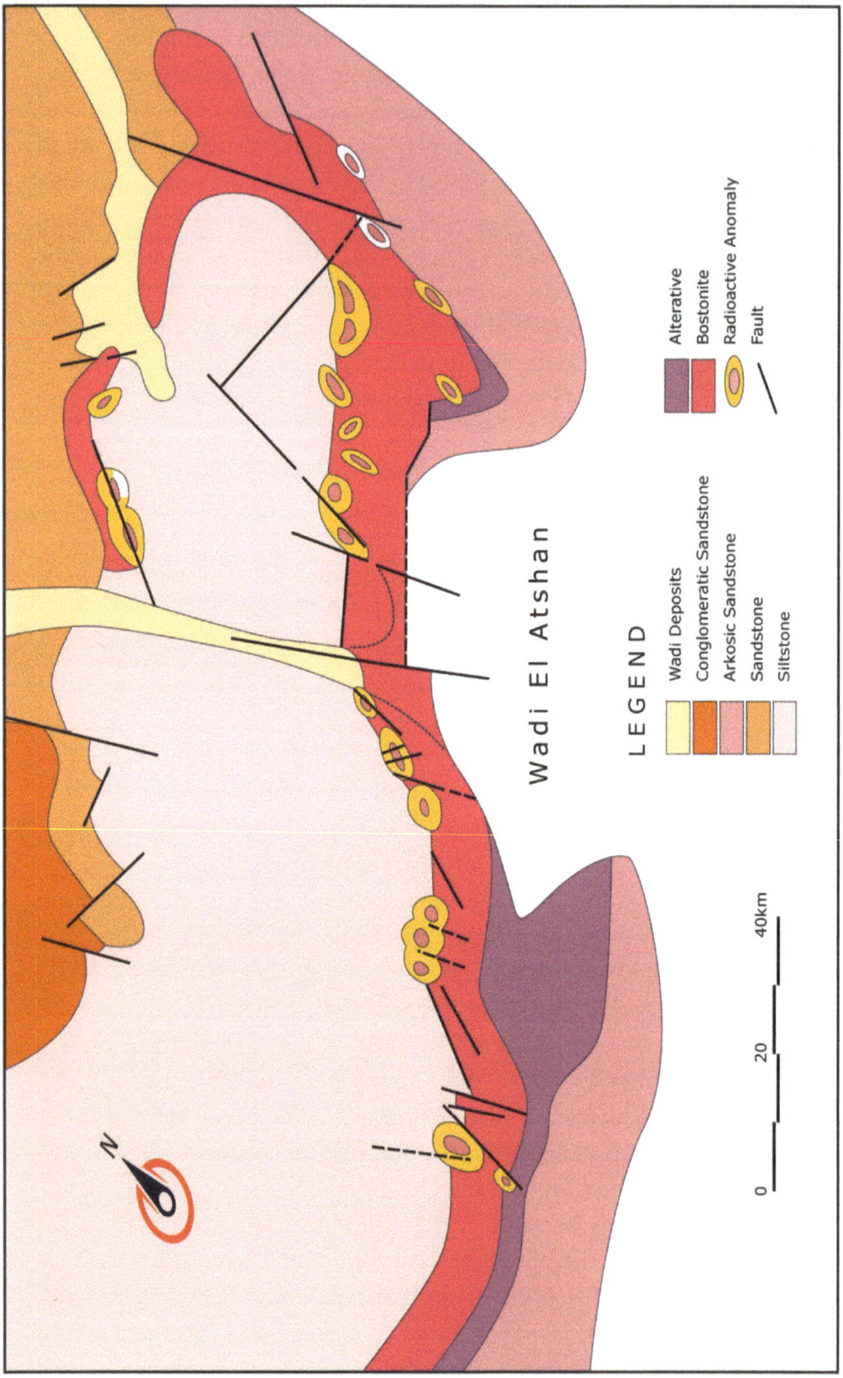

Fig. 5.3.16 ***Geological and radiometric map showing bostonite dyke in contact with Hammamat sediments, showing numerous radioactivity anomalies Central Eastern Desert, Egypt.***

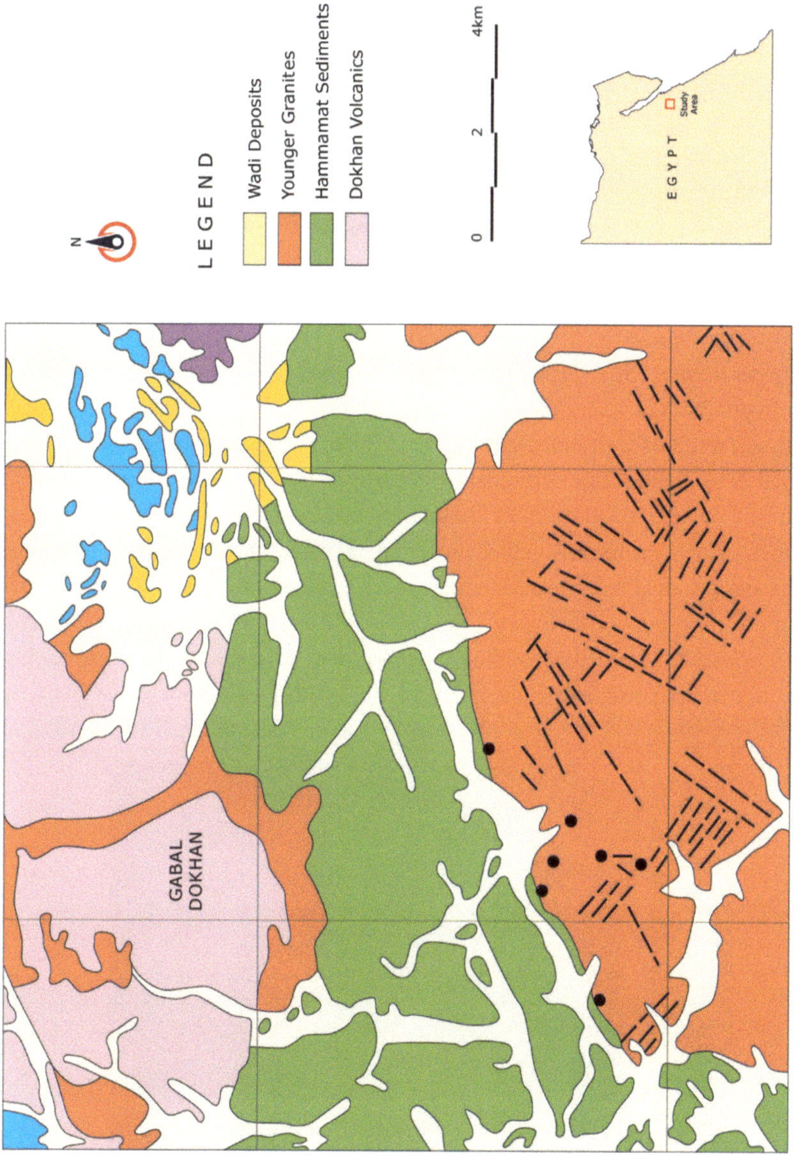

Fig. 5.3.17 ***Um Tawat intra-cratonic basin, Jebel Qattar area, North Eastern Desert.***

between uranium content and the reducing components particularly sulfur (pyrite). These analyses have shown some traces of primary uranium mineral (uraninite). The analyses indicate the importance of Gabal Um Tawat intra-mountain basin as a potential target for economic uranium deposits. The results showed that the basin comprises the reducing conditions and promising structural and lithological features that show its value as a potential uranium site (Salman et al., 1990).

The South Jebel Urn Tawat basin is in the north Eastern Desert to the west of Hurghada City (Red Sea). The basin contains a thick succession of the Hammamat sediments forming an inter-mountain basin. These sediments lie over younger granites that are occasionally exposed on the surface. The southern zone of the basin contains the Gebel Qattar uraniferous granite. Secondary uranium mineralization has been found within the Hammamat sediments at the contact zone with Gebel Qattar granite (Figs. 5.3.18–5.3.20). Along this zone, hydrothermal alterations including hematitization, kaolinization, carbonitization, silicification, epyseynitization, and fluoritization have been found (Salman et al., 1990). The secondary uranium mineralizations are being found within the episeynitized granite. These alteration zones often support uranium mineralization within this province. To determine the economic potential of this region detailed ground geophysical exploration is needed to delineate the configuration, structural framework, alteration zones including any sulfide bodies within this province. The South Wadi El Kareem Hammamat Basin has a semicircular shape with a dimension of about 50 × 50 km.

Fig. 5.3.18 ***Hammamat-Granite contact, Q-V, Wadi Bali.***

Fig. 5.3.19 *Epyseynitization at the contact zone between Hammamat and J. Qattar granite.*

Fig. 5.3.20 *Altered Hammamat sediments with secondary uranium mineralization at the contact zone with Jebel Qattar, North Eastern Desert, Egypt.*

The basin is filled with Hammamat sediments where the sediments have been slightly metamorphosed and are of Late Proterozoic in age (Fig. 5.3.16). The importance of this basin is mainly due to its large size and the presence of numerous igneous lithologies within it. It is also surrounded by younger granite exposures along its northern and southern boundaries. The basement rocks of this region and all Proterozoic regions have a great potential for hosting uranium resources. Table 5.3.2 shows the classification and characterization of the Proterozoic rocks within Egypt and the uranium-bearing units.

Table 5.3.2 The main characteristics of the bostonite generations, Central Eastern Desert, Egypt (After Salman, 1983).

Age (my)	Type of mineralization	Radioactivity distribution	Strike	Mode of occurrences	Locality
74	Not detected	No anomalies	N75°W–S75°E	First generation	Wadi El Kereim
117	Thorium–uranium	Disseminated	N55°E–S55°W	Second generation	
300	Mainly uranium	Vein type	N85°E–S85°W	Third generation	
273	Thorium–uranium	Vein type	N26°E–S26°W	Bostonite sills	Wadi El Atshan
129	Thorium–uranium	Disseminated			
			ENE-WSW	Mainly bostinate sills	Jebel Nasb El Qash
245	Mainly uranium	And vein type			
					Jebel Um Shagher
273	Uranium	Vein type	E-W	Elongated mass	
Not analyzed	Thorium–uranium	Disseminated	N60°W–S60°E	Second generation dykes	Um El Khors and Wadi Farkha
				Firs generation dykes and sills	
302	Mainly uranium	Vein type	NW-SE		

Basal Paleozoic conglomerate

The basal Paleozoic conglomerate units are considered to be one of the most productive potential uranium locations in Egypt. Radiometric anomalies were detected in parts of the Western Desert, Eastern Desert, and Sinai. One of these anomalies will be discussed next. The Jebel El Zeit anomaly within the North Eastern Desert is found along unconformity that is between the Precambrian basement rock and the Nubian sandstones (NEDDD, 1998). The thickness of this highly radioactive zone ranges from 0.2 m to more than 1 m. It is characterized by high degrees of hematitization. This zone is mainly composed of sub-rounded quartz pebbles embedded in a hematitized sandy matrix (Salman et al., 1990).

Chemical analyses of samples from Jebel El Zeit basal Paleozoic conglomerate unit indicate that the uranium and thorium contents range from 181 to 542 ppm and 23 to 174 ppm, respectively. One of the analyzed samples contains 23.75% $Fe_2 O_3$ and 181 ppm uranium. The amount of the leachable uranium in this sample is 109 ppm, which is about 60% of the total uranium. The average uranium content in the other samples is 493 ppm, while the average leachable uranium is 30 ppm. These results indicate the mode of the heterogeneous occurrences of uranium and thorium in the unconformity zone of Jebel El Zeit radioactive anomaly. Some of this uranium is adsorbed by the iron, while the remaining is often present within the structure of the refractory minerals (Salman et al., 1990).

Paleozoic sedimentary sequence

The Paleozoic exposures in Egypt are of a limited distribution. They outcrop next to Precambrian basement rocks in the south Western-Desert, Eastern-Desert, and Sinai areas. The uranium distribution in the Paleozoic rocks and associated provinces of Egypt are good targets for uranium exploration. The Nubian sandstones located near uranium-rich granites and volcanic lithologies can be considered as a favorable host rock for uranium mineralization (Fig. 5.3.21).

Gebel Al Uwaynat area

The Paleozoic strata in the Gebel Al Uwaynat area are located between Gebel Al Uwaynat and the Abu Ras Plateau west of Gilf Kebir. These strata of Ordovician, Silurian, Devonian, Carboniferous, and Permian to Triassic age with a combined thickness of up to 1000 m are present at the eastern edge of Al Kufra Basin and the south west of Dakhla Basin (Klitzsch E., 1990).

Fig. 5.3.21 ***Nubian sandstone exposure to the west of El Missikat uranium prospect, Eastern Desert, Egypt.***

In this province (El Aassy, 1984), there are radioactive anomalies and high uranium and/or thorium content. Uranium anomalies are found in the Silurian Tanezzuft, Devonian ouan Kasa, Carboniferous Assedjefar, and Dembaba, and in the Zarzaitine formations in the western part of Murzuq basin (Almehdi et al., 1991). Precambrian rocks with younger granite intrusive are present on its south east side. The South Western Desert of Egypt includes some structural basins following an ENE-WSW and NNW-SSE trend. The Nubian sandstones in the south Western Desert include the Lingula shale unit and plant bed unit. So, this area includes source rocks and other favorable characteristics to host uranium resources.

Um Kharit rift basin

The Um Kharit rift basin (Nagati, 1986) is located to the southeast of Aswan. It is almost 150 km in length and 40 km in width. This basin has a thick succession of Paleozoic and Mesozoic sediments. This rift is surrounded on the west, east, and south by Precambrian basement rocks that include post orogenic granites. These granites have several radiometric anomalies and secondary and some primary uranium mineralization (Abdel Meguid, 1986). The mode of occurrence of this province and the associated geological and metalogenic environments increase its potentiality for identification of uranium deposits.

Um Bogma

The Um Bogma area is situated in the West Central Sinai and is mainly covered with Paleozoic sediments. The uranium mineralization in this area is found in several of the Lower Carboniferous sedimentary units. The Lower Carboniferous units are subdivided into lower sandstone, middle dolomite and limestone, and an upper sandstone. The middle dolomite unit consists of brown to gray dolomite, limestone, shale, silt, claystone, and sandstone. The shale is varicolored and contains organic matters of varying concentrations, while the sandstone is ferruginated and highly argillaceous and claystone is ferruginated and contains organic materials. Green copper minerals were found in the siltstone as filling fissures and cavities. Siltstone clay and sandstone showed radioactive anomalies incorporated between the two carbonate horizons.

The manganese-iron–rich ore horizon is located at the base of the Lower Carboniferous sequence. The uranium concentration in the sandstone averaged from 105 ppm to 438 ppm, while in the silty strata, it averaged from 955 ppm to 5083 ppm (El Kattan and El Aassy, 1992). The basement units are thought to be uranium source areas and the hydrothermal operations are expected to have beneficial effects; in addition to the presence of oil fields in the region, higher Fe content with high organic material in some Paleozoic units with high fractures and fault densities makes this site suitable for deposition and redistribution of uranium.

Phosphate deposits

The phosphate deposits in Egypt are classified into three sectors: northern, central, and southern. The northern sector includes El Bahariya, Sinai, and parts in the northern Eastern Desert. The southern limit of this division is to the north of Qena. In Sinai along Wadi Sudr, in the flanks of Jebel Zeit along the western coastal zone of Gulf of Suez, and in Baharia Oasis phosphatic beds are associated with pre-Campanian chalk beds and are found in shallow water Santonian sands and shales (Salman, 1974). Phosphate bearing sediments in Egypt is shown in Fig. 5.3.22. Some phosphate deposits were

Fig. 5.3.22 ***Phosphate-bearing sediments in Egypt (modified after Spanderashvilli and Mansour, 1970).***

found in the Arief El Naqa at eastern Sinai. The central sector includes the phosphate-bearing sediments in the Western Desert including the Dakhla; Kharga Oases, Nile Valley; and the Quseir-Safaga region along the Red Sea. These sediments are Campanian and Early Maestrichtian. In this sector, the main economic phosphate deposits of Egypt were found.

In the Western Desert, a huge phosphorite deposit has been discovered at Abu Tartur Plateau, about 50 km west of Kharga Town, the detailed works covered an area of 12 km^2. These phosphorites occur in terrigeneous-carbonate sediments of Maestrichtian age and include one bed with a 3.87 m average thickness. The average P_2O_5 content is 25.65% with 987.8 million tons ore reserves (Wassef, 2003). The Rare Earth element (REE) concentration in the Egyptian phosphate samples from the Abu Tartur deposit was between 0.8 mg/g and 2 mg/g relative to a larger number of phosphate deposits in other parts of the world (Eid et al.,1995), which is remarkably high. The Nile Valley phosphate deposits belong to the central phosphate deposits resources in Egypt. In this region, the phosphate-bearing sediments form a belt that has a 2–25 km width and extends about 170 km (latitudes 25°N and 26°30' N). The phosphate-bearing sediments belong mainly to the Campanian stage and range in thickness from 60 to 90 m (Ghanem et al., 1970). The phosphate-bearing sediments showed vertical facial changes with the lower part being characterized by the predominance of clays and sands, the middle part being marly limestone with chert association are characteristic, while the upper part is chalky, marlym and clayey. The phenomenon of increasing carbonate-bearing rocks upward indicates transgressive environment. The uranium content in the East Luxor phosphate deposits (Nile Valley) ranges from less than 50 ppm and 270 ppm. The high uranium contents are not always associated with high P_2O_5 concentrations (Salman, 1974). The total reserve of the Wadi El Mishash (Nile Valley) phosphate deposits is 359 million tons (Ghanem, et al., 1970) with 22.67% average P_2O_5 content (C_2 and possible category).

The Quseir-Safaga phosphates deposits also belong to the Egyptian central phosphate sector. The phosphate deposits in this area include all the old Egyptian phosphate mines. Almost all these mines are closed now due to either decreasing P_2O_5 grade and/or technical and mining difficulties. In the southern sector to the south of Kom Ombo, along the Nile Valley, till Dungul Oasis in the south Western Desert, the phosphate beds are poorly developed, but associated with lime beds that are equivalent to the chalk of Late Upper Paleocene age or even lower Eocene age, determined by the presence of Osterea Osiris (In Said, 1962). It is clear from the above distribution of phosphate deposits in Egypt, that a large proportion of these deposits are concentrated in the Nile Valley and Abu Tartur Plateau in the Western Desert. The uranium contents of the Nile Valley phosphate deposits are the highest where they range from 50 ppm and 270 ppm (Salman, 1974). The uranium contents in the Abu Tartur Plateau phosphate deposits are relatively low where they range from 20 to 50 ppm. However, these deposits are rather rich in the REE. Therefore, the phosphate deposits in Egypt could be an additional resource for uranium and REE. The uranium and REE can be extracted as byproducts.

To accomplish the above will necessitate the construction of facilities for phosphoric acid purification and the extraction of uranium, REE, and other elements. These facilities should be constructed at the same site where the phosphoric acid and fertilizers production factories are located. It is also recommended that the new phosphoric acid and fertilizers production factories should be constructed nearby the phosphate resources at the Nile Valley and Abu Tartur Plateau. Most of these factories are now located in Nearby Cairo and Nile Delta, which are more than 1000 km far from the resources. Additionally, black shales are present in large quantities in Egypt mainly in Upper Cretaceous-Lower Tertiary sedimentary units. They are mainly close to the phosphate deposits especially in the main economic phosphate sector in Egypt (central sector). Some units of the black shales are rich in heavy metals especially vanadium, zinc, and cadmium besides uranium. The black shales cap the economic phosphate-beds in several locations in the Quseir area and they are enriched in vanadium (0.45%), zinc (0.4%), and other heavy metals. Moreover, this black shale is fertile in uranium (El-Kammar et al., 2009). This indicates the importance of the black shales in the localization of potential uranium resources in Egypt. It is of prime importance to study in details the distribution of the black shales and the associated structural features, especially the basin structures that can act as good traps for uranium, if the geochemical features are suitable.

Placer deposits

Egypt has a large amount of place deposits that are mainly located in the north Nile Delta and northern Sinai along the Mediterranean coastal stretch. These deposits include economic minerals that are essential in the nuclear industry and metallurgical and engineering industries. The economic minerals are monazite, zircon, rutile, ilmenite, garnet, and magnetite. The monazite contains REEs in the mineral—thorium anhydrous orthophosphate with assays of 0.48% U_3O_8, 6.04% ThO_2, and 61.31% RE_2O_3. The zircon is in the form of zirconium silicate mineral and contains 67.06% $(Zr+Hf)O_2$. Traces of uranium range from 0.04% to 0.11% U_3O_8, while its thorium content varies from 0.02% to 1.19% ThO_2. The rutile is almost a pure titanium dioxide assaying up to 98.64% TiO_2 (Dabbour, 1994). These Egyptian placer deposits extend about 500 km from Rosetta to Rafah. They are present in two forms: beach black sand sheets and sand dunes along the Mediterranean coastal zone. This region contains a large amount of the placer deposits that can represent an important supply for uranium minerals, thorium, and REE in Egypt.

CHAPTER 5.4

Islamic Republic of Iran

Introduction

Iran is located in southwestern Asia and northeastern Arabia. The country is bounded by the (formerly) Soviet Union and the Caspian Sea from the north, and Afghanistan and Pakistan to the east. The Arabian Gulf and the Gulf of Oman are to the south, and Iraq and Turkey border Iran to the west. Iran covers an area of 1,633,188 km^2. The terrain can be divided into four main regions: (1) The inner plateau, (2) mountains, (3) the Caspian Sea coast, and (4) Khuzestan Plain. The inner plateau is located in the center and west of Iran and occupies about half of the total area of the country. The plateau rises about 900 m above sea level, and is surrounded by mountains to a large degree. Most of the surface of the inner plateau covers two large, almost uninhabited deserts, the Great Saline Desert and the Lot Desert, which are among the world's driest and most arid deserts, and together occupy more than 98,000 km^2. Two large mountain ranges, Elborz and Zagros, border most of the inner plateau. The Elbrus mountains are located along the northern border of Iran. They form a massive wall between the Caspian coast region and the inner plateau. The highest peak in Iran—Mount Damavand—rises 5,604 m above sea level. A large amount of rain falls on the northern slopes of Abruzzo, and farmers cultivate a variety of crops on fertile land. The slopes of the southern chain are relatively arid and dry.

Geology of Iran

Geologically, the Iranian plateau region consists of growing Bondan tectonic fragments between the Turan Wharf to the north and the main Zagros crust, the fusion area between the Arabian plate to the north and the Eurasian plate. Throughout Western Asia and Central Asia, the Iranian Plateau or the Persian Plateau is a geological topography. The plate is partly the Eurasian plateau, jammed between the Arab and Indian plateau, situated between the Zagros Mountains in the west, Kopet Dag in the north, and the Caspian Sea in the northwest, the Armenian highlands and the Caucasus Mountains in the northwest, the Strait of Hormuz and the Persian Gulf to the south and the Indus River to the east in Pakistan Fig. 5.4.1. The Zagros Mountains form the western edge of the plateau and may include its eastern slopes. The Iranian Plateau stretches for about 2000 km from the Caspian Sea in the northwest to Baluchistan to the southeast. In an area west of the Indus river, the largest part of Iran, Afghanistan and Pakistan form an area that takes on an almost quadrangular shape consisting of the city of Tabriz, Shiraz, Peshawar, and Quetta with an area of 3,700,000 km^2. Although it is called "plateau," it is far from flat terrain but contains many mountain ranges, where Mount Damavand in Alborz is the highest point at 5610 m, and the lot of East Kerman in central Iran is its lowest points down 300 m.

Iran is composed of several geological units ranging from Precambrian to Cenozoic in age. The major geological units in Iran are mainly controlled by the tectonic episodes that have occurred since the Precambrian. The Iranian basement has a mixture of Proterozoic lithologies formed during the Baikalian or Pan-African orogeny and consists of a series of structurally complex metamorphic and granitic lithologies. The Iranian basement rocks are formed between 600 and 1100 Ma and are similar to Arabian Shield rocks. An important event that took place during this period is the post-Pan-African magmatism as evident by the Doran Granite. In Iran, the Late Precambrian (Eocambrian) postorogenic volcanic event contains rhyolitic tuff, alkali rhyolite, and basic dykes known. The basement consists of greenschist facies, fine-clastic sediments. This portion of moderately metamorphosed Precambrian basement has been the subject to a questionable unconformity with overlying non-metamorphosed Eocambrian-Cambrian age sedimentary rocks. The deposited series of rocks reveals all the features of a platform cover. It suggests an epicontinental climate with alternating shallow-marine, lagoonal and continental deposits. Evaporites (salt and gypsum) collected in the domes and ridges of the larger anticlinal structures after pan-African orogeny in southern Iran and the southeastern part of central Iran were deposited.

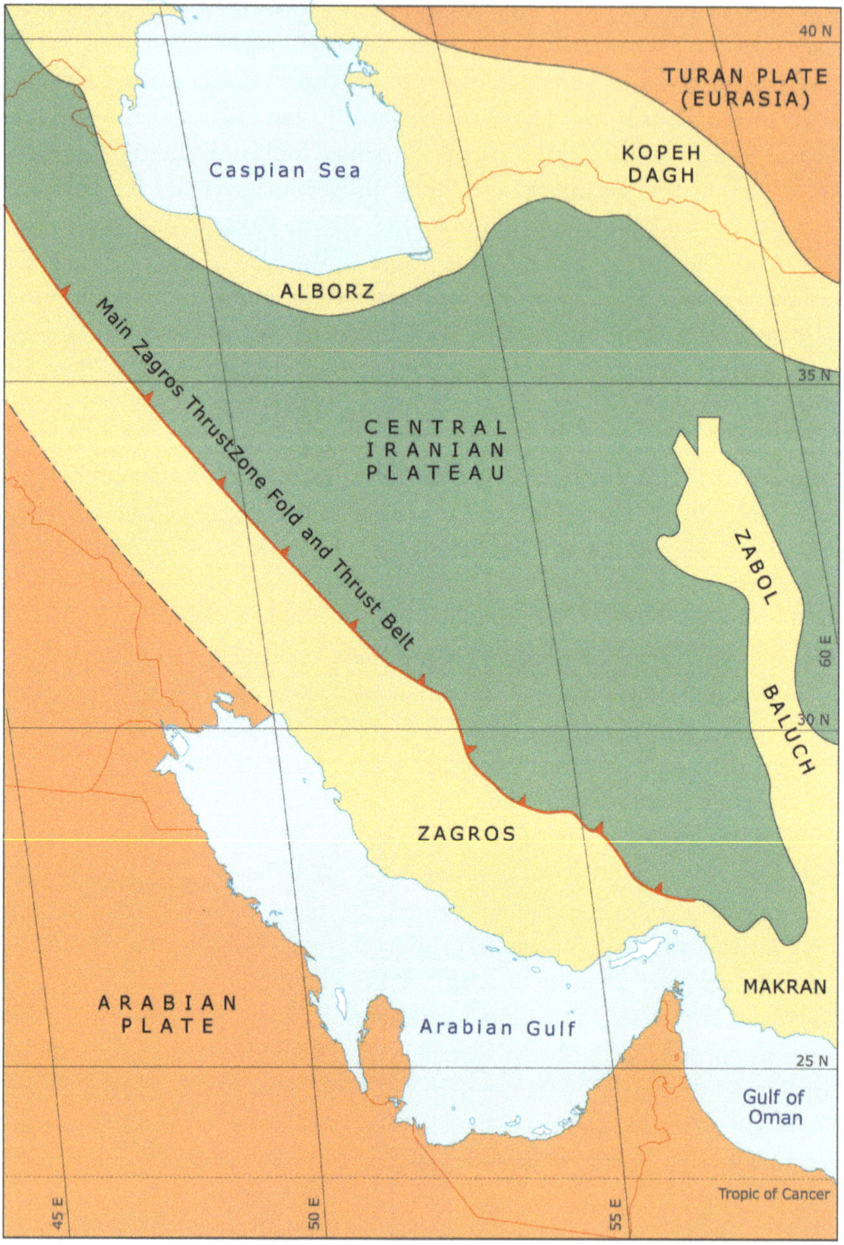

Fig. 5.4.1 ***Map showing Zagros folded and thrust belt and Iranian Plateau.***

Uranium provinces

There is scant information about the uranium resources in Iran, especially the two known main uranium provinces (Fig. 5.4.2). These two provinces are the Central Iran and Azarbayjan uranium provinces. The Central Iran Province includes the following uranium prospects: Anarak, Ariz, Chah Juieh, Esfordi, Khoshoumi, Lakeh-Siah, Narigan, Saghan, and Zarigan. The Azarbayjan Province includes Mangar Abad, Maragheh, Talkhehround, and Uroumiyeh uranium prospects. There are additional uranium prospects in the Bandar Abbas region in Iran (Saeidi, 2005).

Saghand uranium deposit

The Saghand uranium ore deposit located within the Central Iran Province includes the Esfordi, Lakeh-Siah, Aiez and Chapendomi, Rig-e-Zarin, and

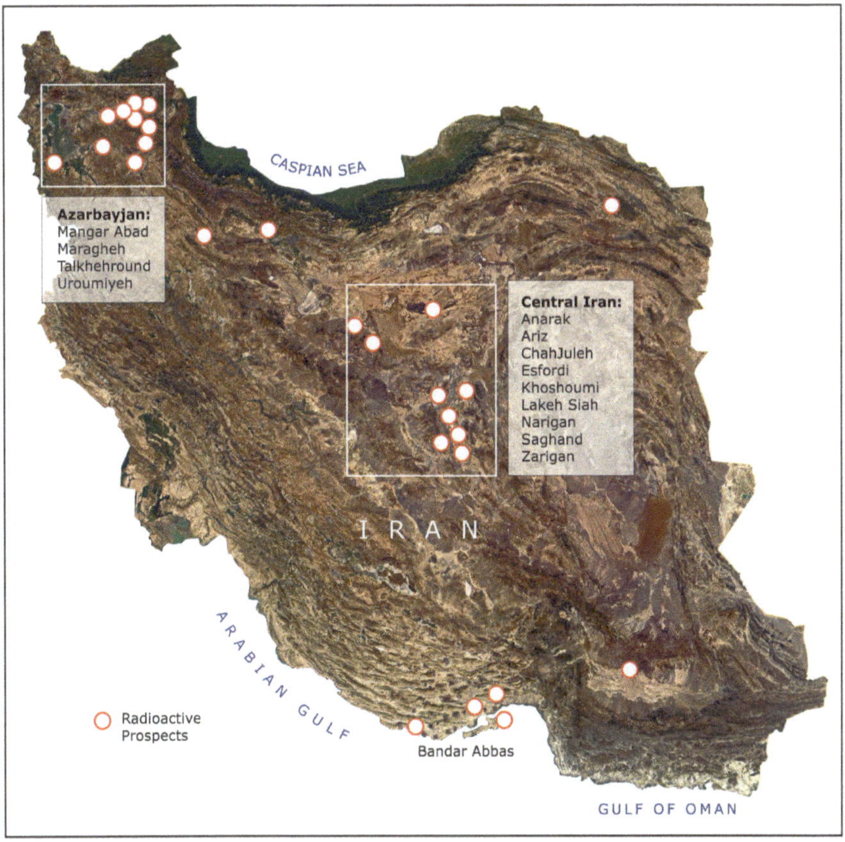

Fig. 5.4.2 ***Radioactive prospects in Iran (After: Saeidi, M., 2005).***

Natk prospects. Exploratory activities within the Saghand area (Fig. 5.4.2) commenced after the Islamic revolution by utilizing airborne geophysical surveys. Much of the activity was at the early exploration level until 1989 and the total area explored was 200 hectares. The region covering significant radioactivity anomalies was given priorities for more systematic exploration beginning in 1990 (Fig. 5.4.2). The Hungarian ELG company used logging techniques and the lithological logs provided by the geologists of Iran and China to determine a 350 m deep region where the ore reserve calculation was made. Iran announced in February 2003 that it had started mining deposits of economic uranium at Saghand near Yazd. The uranium ore deposit measures 100–150 square kilometers, with approximate reserves of 3000–5000 tons of uranium oxide.

The mine project covers an area of 20 hectares, 185 km northeast of Yazd, The exploration was completed in 1994 and found areas with an estimated reserve of 1.58 million metric tons of uranium ore, with an average grade of 533 ppm (0.0533% uranium) which represents 842 metric tons of total uranium content. The final production will include two shafts in an underground mine, each 350 m deep. Ore grades greater than 300 ppm will be shipped to a uranium mill, while ores of grades from 100 and 300 ppm will be used for heap leaching. At the interval between 270 m and 400 m, the uranium mineralization is found in lensoid bodies that slope between 5° and 30°. The size of the deposit is 300 by 400 m. The ore reserve has been estimated to be 1,398,000 tons with an average grade of 562 ppm in two categories: C and D. Dispersed lenses with a slope of 50–70° degrees to a 16 m depth start mineralizing in deposit No.1 from the ground level. The size of the deposit is 200 by 300 m. The deposit of ore reserve is estimated as 152,000 tons with the average grade of 459 ppm according to two grades: C and D.

The exploration and evaluation of uranium resources in the Bafq-Posh- e-Bad metallogenic zone, contains the Saghand, Narigan, Zarigan, Chah-Juueh, Khoshomi, Sfordi, Lakeh-Siah and Sechahun deposits. These uranium deposits are found in Upper Precambrian magmatic and metasomatic lithologies. The most important uranium mineralization in Narigan is hosted within Precambrian alkaline granite. At the Zarigan deposit, thorium and uranium anomalies are, respectively, to the north and south of the deposit (is this what you mean, it is confusing what you mean here). There is the potential for uranium resources within the albite metasomatic rocks in the Chah-Juueh area, the rugged topography restricts its exploitation, but in general it is favorable for future production (Uranium, 2005).

The Azarbayjan Province includes Mangar Abad, Maragheh, Talkhehrund, and Uroumiyeh uranium prospects. The uranium resources in this province are characterized by hydrothermal polymetallic deposits (Uranium, 2003).

Uranium resources potentialities in Iran

The Doran leucogranite

This site represents an important post Pan-African magmatic event, which can be considered as a source and host rock for uranium resources. The fault zones, associated fractures, and their intersections have been found to be good mineralization traps. The associated alkali rhyolites and the rhyolitic tuffs are a potential source and host rocks for uranium. The presence of some basic dykes in these post orogenic granites and rhyolites is important in localizing the deposits. The basic dykes form barriers in the leucogranites and rhyolite, which act as entrapments for localizing uranium mineralization. Also, the intra-cratonic basins in Iran, which are filled with post orogenic low-grade metamorphosed clastics units, can be considered good targets for potential uranium resources in northern and eastern Iran.

Palaeozoic sequences

The Paleozoic sedimentary sequences are found in the main and south Alborz, the majority of central and eastern Iran, and in southwestern Iran. The Palaeozoic sedimentary units represent one of the important targets from uranium deposits in Iran. The basal conglomerates and the sandstones units associated with some shale and silt beds, in addition to some organic remains are favorable for hosting uranium resources. The Lower Paleozoic sequences are equivalent to the Nubian sandstone facies that is well developed in Egypt, Sudan, and Libya and shows some radiometric anomalies. The basal conglomerates that partly cover basement rocks in the Eastern Desert and Sinai in Egypt showed some important radiometric anomalies (Salman et al., 1990; El Khouly, D. M, 1996). Also, the Upper Carboniferous units have potential for hosting uranium deposits. These include the carboniferous Um Bogma formation in Sinai and the clastic carboniferous beds in Wadi Araba, Eastern Desert, Egypt (Salman and El Aassy, 1983). The Lower Dolomite member of the Soltanieh formation contains an assemblage of phosphatic tubes and other poorly preserved remains (Hamdi and Zhiwen, 1989). The intra-cratonic basins, where thick sequences of Palaeozoic sediments are often deposited can be consider as good target for uranium exploration in central Iran and in the vicinity of the Zagros folding belt. In these basins there are good opportunities for entrapment of the leached and moved uranium from the acidic igneous and volcanic rocks in the nearby basement terrains.

Salt plugs

Evaporites (salt and gypsum) deposits, later formed into domes and crests in larger anticlinal systems, in southern and southeastern portion of central Iran were formed after the Pan-African orogeny. These evaporate deposits were formed during late Precambrian-Cambrian period in central Iran and the Zagros region. There are some potential uranium deposits especially the Bandar Abbas deposit. Several exploration tests have shown that large amounts of uranium ore at the Bandar Abbas deposit are detectable in open pits. The annual output of 30 tons of U_3O_8 seems feasible at lower mining costs compared to Saghand, according to initial estimates.

Phosphate deposits

Phosphate reserves are abundant in Iran, with proven reserves of about 500 million tons. It is enough to satisfy roughly the annual consumption of approximately 2.5 million tons of phosphate fertilizer. There are phosphate reserves in all regions in Iran. It is clear from the above information about the characteristic features of the phosphate deposits in Iran that these deposits can form a source for uranium. The presence of these phosphates in more than one mode of occurrence suggests that they can also be considered a source for uranium. So, it is expected to find some uranium deposits in the nearby areas, if there are structural traps and suitable geochemical environments.

CHAPTER 5.5

Iraq

Geological overview

Iraq is one of the Middle East countries, bordered on the west by Syria and Jordan, by Saudi Arabia on the southwest and by Kuwait from the south, Iran on the east, and Turkey on the north. It is connected to the Persian Gulf on the southeast.

There are four major geographic regions in Iraq, the desert areas in the west-southwest, the rolling-uplands between the Tigris River and Euphrates, mountainous areas in the north-northeastern region, and the alluvial plain where the Tigris and Euphrates rivers flow. The total land area is about 438,446 square kilometers. The desert area is part of the desert of the Syria, heading west to Syria and south to Saudi Arabia. The uplands areas extending between north of the Tigris and Euphrates are called the Island or Al Jazirah. Iraq's northeastern region is highlands that reach to the Iranian and Turkish borders. The region begins as larger steppes and mixes into mountains of between 1000 and 4000 m.

Iraq alluvial plains are located in a structural trough, known as the Mesopotamian Basin (Master and Woldai, 2004), which is linked to active subduction-related orogenic processes in the Zagros Mountains of Iran and northeastern Iraq (Fig. 5.5.1). The basin belongs to the broader Zagros foreland, connected to the closing of the Neotethys Ocean and the Arabian and Eurasian plate collisions. Owing to the convergence of the Zagros collision zone, the area remains tectonically active. Neoproterozoic crystalline rocky

Fig. 5.5.1 ***Map showing Iraq and the adjacent countries.***

basement of the Arabian shield floors the Mesopotamian Basin. A thick layer of Phanerozoic-sedimentary rocks consisting of (1) a Paleozoic sequence of Cambro-Ordovician, Devonian-Lower Carboniferous, and Upper Permian rocks; (2) a Mesozoic succession of Triassic, Jurassic, and Cretaceous rocks; and (3) a Cenozoic succession of Eocene to Pliocene rocks, overlain by Pleistocene to Holocene alluvium over lies Proterozoic metamorphic and igneous basement lithologies. The alluvium, which includes clay, silt, sand, and gravel, is related to the floodplain of the Euphrates and Tigris rivers and associated swamps, as well as to marine incursions.

Uranium resources in Iraq

There is scarcity of information about uranium exploration and resources in Iraq. Prospecting at the Abu Sukhayr—which is 25 km southwest of Najaf (UN, 1991)

> *"carbonate ore mine started in September 1988 and ended with flooding at the end of 1990. During its operations the overall production of marley limestone has been to be 800 tons and its average uranium content is 150 parts per million, 2–3 times the level of ore in Akashat. Geologic Institute in Baghdad was working on a uranium recovery process in Sukhayr" (IAEA, 1993).*

The Iraqi Geological Survey Institute was the site for the production and pilot project of uranium recovery from the carbonate ore (such as uranium UO_4) from Abu-Sukhayr mine (25 km south of Najar). In June 1990, the pilot plant was completed; the Institute received 20 tons of ore from Abu-Sukhayr in July 1990. Ten tons, producing 500 g of UO_4, have been processed and the remaining 10 tons are returned to the mine. The facility is currently used for the extraction of aluminum (IAEA, 1992). Akashat—420 km west of Baghdad—(Fig. 5.5.2) is the key area of Iraq's uranium mining operations. At least 164 tons of yellowcake were produced in Iraq in the mid-1980s, most of it was mined in Akashat and milled at the Al-Qaim plant (Nardulli Giuseppe, 1998). The phosphate open cast mine in Akashat supplies reserves of phosphate rock processed for uranium extraction in Al-Qaim facility (UN, 1991). The annual production of phosphorus pentoxide (P_2O_5) in Iraq at Akashat was estimated to be about 1 million tons during the 1990s with 300,000 tons. In 2001 phosphate rock output of 100,000 tons was obtained from P_2O_5.

Northeast of Mosul, a uranium mine is located, but no further information can be found. During processing of the fertilizer complex west of Al Qaim, uranium was also recovered from phosphate rocks, but quantitative

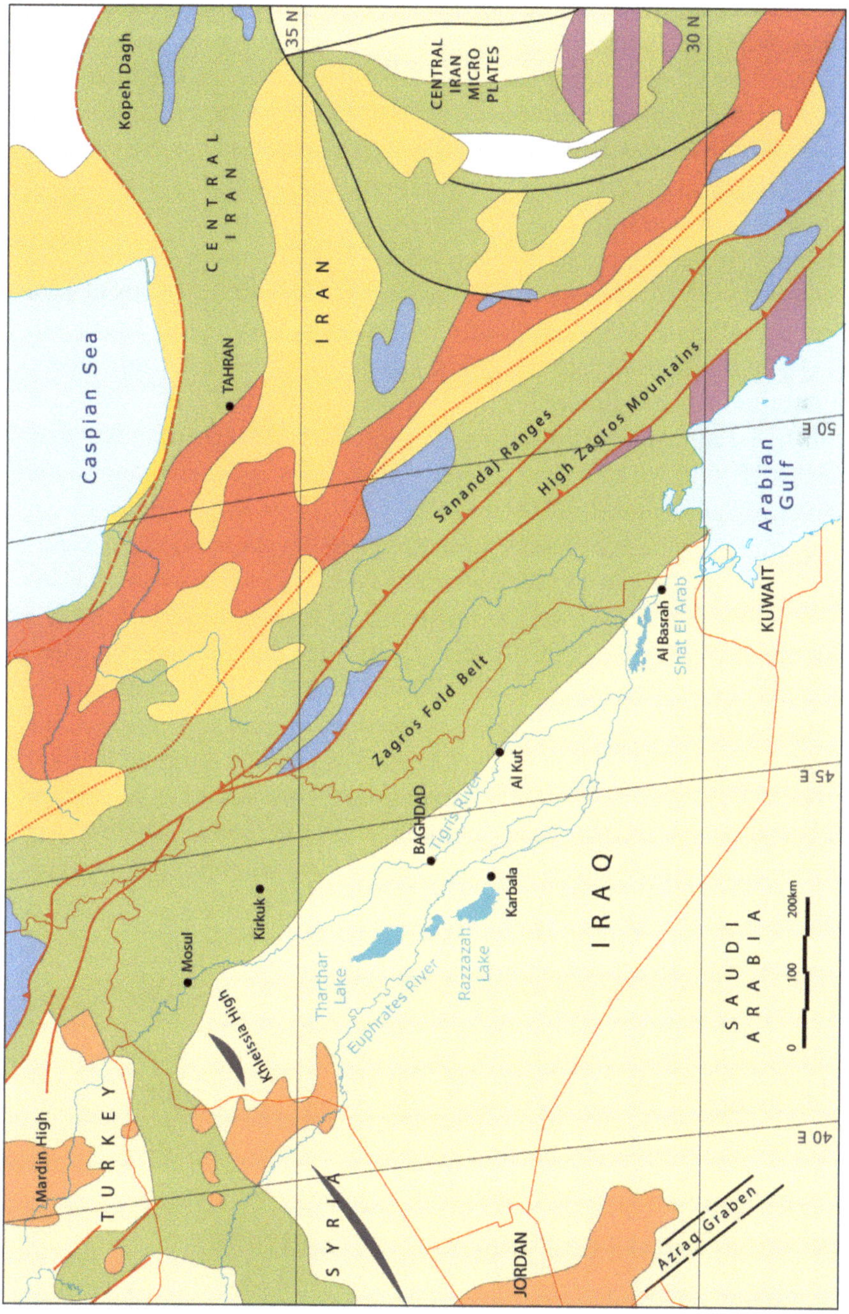

Fig. 5.5.2 *Tectonic map of Zagros fold belt and its extension at the north east of Iraq.*

data were not available. Additionally, Iraq has phosphorite deposits as a part of the major phosphorite belt in North Africa and the Middle East. In western Iraq desert of Tethyan province, bedded marine sedimentary phosphate rocks of Campanian and Maastrichtian era are exposed. The phosphate and associated carbonate rocks are enriched in Cr, Ni, Cu, Zn, V, and organic matter. Isotopically light carbon and heavy sulfur enhance apatite (Al-Bassam et al., 1983). The concentration of uranium in the phosphorites is between 0 to 102 ppm (average 75 ppm). In the Precambrian and Cambrian deposits, the average uranium level is of 4–48 ppm, and it is 20–90 ppm in Palaeozoic and Jurassic deposits, 40–130 ppm in Late Cretaceous–Paleogene deposits, 30–130 ppm in Neogene deposits, and 30–110 ppm in Quaternary (including Holocene) deposits. On the whole, the variation range is similar for phosphorites of different ages.

In Iraq, the uranium/P_2O_5 ratio in phosphorites ranges from less than unity to 24×10^{-4}. The major phosphorite deposits of the world with ore reserves of approximately 250 Gt (or 58 Gt P_2O_5) contain up to 19 Mt of uranium. Uranium is present in phosphorites in the tetra- and hexavalent, that is, U (IV) and U (VI) states, and their ratio is highly variable. At the early diagenetic stage of the formation of marine phosphorites in a reductive environment, U (VI) diffuses from the near-bottom water into sediments. The submicroscopic segregations of uranium minerals (mostly uraninite), which are most likely absorbed by phosphatic material, are therefore reduced and precipitated. Uranium is partially oxidized and lost in weathering in a subaerial environment during the resulting reaction of phosphorites and aerated water. Calcium phosphate and non-phosphatic components are replaced by a more complete crystallization in the catagenesis (Baturin and Kochenov, 2001).

Uranium resources potentiality in Iraq

In spite of the scarcity of the available information about uranium resources in Iraq, the geological and tectonic environments suggest that Iraq may host economic uranium deposits. The most favorable provinces for uranium resources potentialities are discussed next.

Younger granites and acidic volcanics and tuffs

The eastern tectonically active zone is in contact with Zagros folds belts and mountains. The younger granites and the associated, acidic volcanic rocks and tuffs represent a potential source and host rocks for uranium. The contacts of these rocks with Late Proterozoic clastic sedimentary rocks and the associated intra-cratonic basins represent suitable traps for economic uranium deposits.

Basal conglomerates and the Palaeozoic rocks

The Palaeozoic clastic sedimentary sequence and their associated basal conglomerates are found at the contact zones with the late Proterozoic basement rocks. These clastic units can act as hosts for mobile uranium from the younger granite and the acid and alkaline volcanics. These units can host economic uranium deposits if the geochemical and rock constituents are suitable. It is noteworthy to mention that the nearby oil fields and the organic gas emission can have a significant role in fixing the secondary uranium-mineralization as a primary uranium type.

Surfacial uranium deposits

There is a good potential for finding surfacial uranium deposits in some areas in Iraq. The first potential area is to the east and northeast of Iraq where there is the continuation of Zagros Mountains and their slopes. It is important to prospect to find out closed basins in these areas. In these closed basins some surfacial uranium deposits could be detected. Another possibility for finding surfacial uranium deposits can be expected in the northwest trending zone at the west of Iraq, that is, from Najaf to Akashat (Fig. 5.5.2. Iraq 3). The uranium can be formed in this area after leaching from the phosphorite deposits in the northwest of Iraq.

Phosphorite deposits

The phosphorite deposits can be considered as a potential source for uranium in Iraq as it is already used for uranium extraction at Akashat's open-cast phosphate mine. It supplies deposits of phosphate rock which can be processed in the Al-Qaim facility of uranium extraction (UN, 1991). In the 1990s, the annual production of phosphate in Iraq at Akashat was estimated at one million tons with 300,000 tons of phosphorus pentoxide (P_2O_5). It is not clear how Iraq extracted uranium from the phosphorite deposits; was it directly from chemical treatment of the ore or from phosphoric acid through purification processes.

CHAPTER 5.6

Jordan

Geological outlines

Jordan contains lithologies ranging from Precambrian to Quaternary in age (Fig. 5.6.1). Precambrian igneous and metamorphic rocks cover an area of approximately 1800 square kilometers in southern Jordan. These rocks are exposed from the area south of Aqaba for approximately 70 km toward north along the east side of Wadi Araba Rift. They are also exposed east of the Rift and south of the Ras en Neqb escarpment, dipping east of Wadi Rum underneath thick Palaeozoic sandstones.

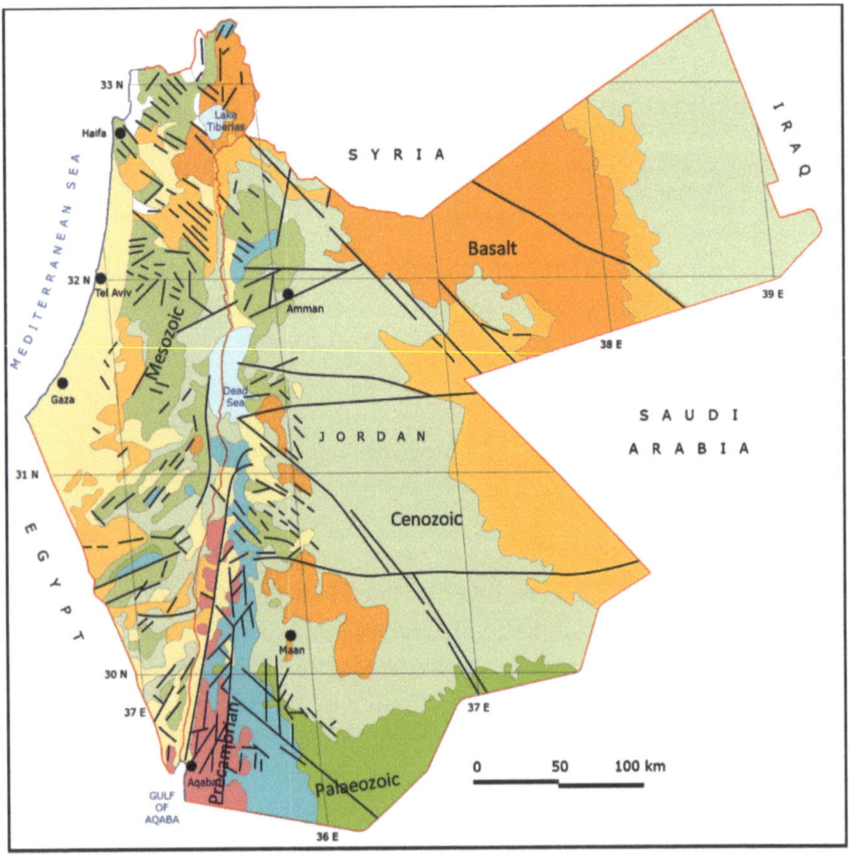

Fig. 5.6.1 *General geological map of Jordan.*

The basement complex is composed mostly of biotite aplite-granite, two mica-aplite granite, granodiorite, quartz-diorite, alaskite, and diorite. These rocks have been intruded by numerous acidic, intermediate, and basic dykes. Based on correlation with similar igneous lithologies in the Sinai desert and the Red Sea area, the Jordanian basement complex is thought to be around 590 million years old.

Uplift and erosion preceded the deposition of the Upper Algoncian sediments that overlie the older rocks with an angular unconformity. These sediments, the Saramuj Conglomerates and the Slate Greywacke Series, have been metamorphosed to low-grade facies. The entire Precambrian basement complex is considered to belong to the northern rim of the African craton.

The main tectonic events in Jordan since the Cambrian are: (1) The Tethys Ocean, which occupied the area in western and northwestern Jordan, and transgressed several times within Jordan; (2) the Wadi Araba-Jordan suture between the Palestine and East Jordan blocks; and (3) the Nubian-Arabian Shield in southeastern and southern Jordan.

During Lower to Middle Cambrian, the first transgression of the Tethys Sea reached the eastern edge of the Central Wadi Araba. In eastern and southeastern Jordan, limit of the transgression continued into an epeirogenic basin during a dominantly dry climate period. In the Middle Cambrian, a regression deposited continental sandstones in northern Wadi Araba, on the top of the marine sandstones. In the Lower Ordovician, the ocean transgressed and covered the entire SE of Jordan. Marine shallow water sediments dominant by fine sandstones and sandy shales were deposited during the Middle and Upper Ordovician until the Liandoverian. During the Upper Silurian continental deposition of the clastics resumed, as observed in the southeastern corner of the country. Lower Devonian sediments equivalent to the Tabuk formation of Saudi Arabia may exist beneath the Mesozoic sediments in the Jafr, Wadi Sirhan, and Azraq Troughs. Palaeontological studies carried out by Olexcon (1967) on material collected from the Safra Well (44 km east of Amman) indicate that sedimentation persisted at least locally in the area into the Carboniferous period. Anyhow it is believed again that remnants of the Carboniferous and Permian sediments might be preserved in the Jafr, Azraq, Wadi Sirhan troughs and in the northwestern side of the country. During the Permian, the Palaeozoic rocks were tilted eastwards and were eroded. Sedimentation resumed during the Lower Triassic, here outcrops are found in the lower reaches of wadis which breach the rift escarpment between the Wadi Hisban and north of El Lisan

Area. Sedimentation during this age was restricted only in the northern and western portions of the country (north of El Lisan area), while further east and southeast erosion prevailed on the continent. During the Triassic phase of sedimentation, a sequence of arenaceous, argillaceous, and carbonate deposits accumulated.

In the Middle Jurassic, the sea regressed again. Outcrops of Jurassic sediments are restricted to the Jordan Valley side wadis in a region extending southwards for about 20 km from Wadi Zerqa. During this period, sandstones, clays, and dolomitic limestones accumulated. Lower Cretaceous sandstones are neritic to littoral north and northwest of the Jurassic shoreline. South and southeast of this shoreline, the lower part of the Lower Cretaceous succession, both marine and brackish influences have been encountered in Southeast Jordan. The Lower Cretaceous sandstones of the continental origin extend to the east and to the extreme southeast where they become diachronous and their top becomes younger from west to east. The Middle Cretaceous (Cenomanian–Turonian) sediments are fully marine and thick in the western, northwestern, and northern Jordan. On the contrary in the south, southeast, and east, they wedge out, calcareous contents decrease, and the arenaceous facies prevail.

The Upper Cretaceous (Santonian–Maastrichtian) marine sediments have been developed into two different characteristic sedimentary successions: (1) The chalky limestone, sandy limestone, silicified limestone, sandstones, chert, phosphate, and Tripoli deposits were originated in what is called "Gentle Swells System" and (2) the thick bituminized marls. Marly limestone, compacted dolomites, and limestones interbedded with black brown chert layers were deposited in basins and troughs with local euxinic environments. The Paleocene Tertiary marine sediments and the majority of the Eocene have been deposited in a similar "smooth swells and basin system." Along the Wadi Araba–Jordan rift, coarse clastic bordering weathering materials were simultaneously deposited in the Oligocene along with taphrogenic tectonism. Either freshwater lakes draining west to the Tethys occupied the Jordan rift during the Cainozoic, or it created an inland depression with terrestrial or lacustrine sediments. Also, marine deposits during the Miocene extended for 100 km east of the rift into El Azraq–Wadi Sirhan depressions and even east of the Jebel Ed-Drouz area. A major phase of taphrogenic movements took place in the plio-Pleistocene. The major graben movements ended at the time of the Middle Pleistocene basalt volcanism. A probable freshwater lake (Samra Lake) covered the northern part of the rift. This lake slowly changed to the brackish

Lisan Lake, and Lisan Maris extends from approximately 50 km south of the Dead Sea to the north almost as far as Lake Tiberias. Not before the Upper Pleistocene did the Lisan Lake change into the present Dead Sea. Throughout the Quaternary, detritus was transported into the Rift Valley, and the extension Azraq-Wadi Sirhan and El-Jafr depressions of East Jordan.

Historical review

The following are the air-based surveys were conducted for the entire country:

1. 1988: A ground radiometric survey of the airborne discovered anomalies completed.
2. 1988–1990: Evaluation of the Precambrian basement and Ordovician sandstone target areas were evaluated
3. 1990–1992: Regional geochemical sampling program for stream sediments and some rock samples was completed for the basement rocks followed by geological and radiometric survey for basement rocks and associated sandstones.
4. 1997: Completed a study and evaluation of uranium concentration in Jordan's phosphate deposits to assess the environmental effects of the uranium.

Radiometric measurements (gamma and radon) and chemical analyses defined several surfacial uranium occurrences in central (related to varicolored marble covering an area of 350 km^2), southern, and southeastern Jordan (Uranium, 2005) Uranium occurs as tiny mineral grains distributed in fine calcareous Pleistocene sediments and as yellowish carnotite and other uranium minerals film coating fractures of Maastrichtian-Paleocene age fragmentary chalk or marl sequence. Uranium only appears as yellowish stains with chalk and marl in the south and southwest zone. The series of the chalk marl in the region investigated is the principal component of the uranium-bearing rocks. The amount of calcite and clay content are low. Uranium content ranging from 140 ppm to 2200 ppm over an averaging thickness of about 1.4 m is shown by the results of channel sampling in three regions in central Jordan. The median overload is approximately 0.5 m.

Uranium resources

From potential uranium resources point of view, parts of Jordan are of interest for exploration with a high potential for hosting an economic uranium deposits. In such areas, the several following criteria for hosting uranium are recognized:

- Biotite aplite-granites and two mica-aplite granites which often represent younger Pan-African granites. These rocks can be considered as source and host rocks for uranium ores as recorded in several areas in the Middle East countries.
- Thick basal sandstone sequences intercalated with shale and carbonate beds represent a suitable environment for hosting uranium deposits.
- Bituminous shale beds, seepages, and asphalt rocks can represent reducing environments which can play an important role in fixing and depositing primary uranium mineralization.
- Phosphate deposits of the Upper Cretaceous age: About 60% of Jordan area is covered with phosphate-bearing formations that are often exposed the near surface or at various depths. These phosphate beds can be considered as host rock for uranium and/or as sources for uranium.
- Volcanic activities before, during, and after the formation of the Dead Sea Rift, thermal water, and meteoric water often play an important role in redistribution of uranium and depositing remarkable amount of surface uranium deposits at various localities in Jordan.

Jordan recently included in its country reports of the uranium group meetings figures about its uranium reserves estimations. These reserves fall in Reasonably Assured Resources and Inferred Resources (Uranium, 2005). According to the available information and the geological environments in Jordan the following potential uranium sites are described next.

Younger granites and acidic volcanics

It is evident that the younger granites are hosting uranium resources in many countries in Middle East and North African Countries such as Egypt and Algeria (Salman, 2005). Younger granites exposures are present in the southwestern parts of Jordan.

Basal conglomerates and Palaeozoic sediments

The basal conglomerate horizons that are often present at the contact zone between the Precambrian rocks and the Palaeozoic sedimentary sequences could host uranium resources. Some of these horizons include high radioactive anomalies in similar horizons found in Egypt (Salman, 1995, 2005). The Nubian sandstones of the Palaeozoic sediments with a good potential for hosting uranium resources as radioactive anomalies were recorded in the Ordovician Nubian sandstone beds in Jordan (Salman, 2005).

The thorium content in these Ordovician rocks reaches more than 1000 ppm in addition to some uranium content; the leachable amount of this uranium is about 45 ppm. The Ordovician sandstones sequence is located in the southeastern areas especially in the Dobideb and Jebel El Medours areas. The uranium content in the basement granites rocks is about 38 ppm, which can be considered as a source rock for supplying the nearby sandstones by uranium. The presence of basin structures often increases their potentiality for hosting uranium resources.

Recent uranium deposits

In north Jordan the uranium resources are present as the following three main types (Xoubi, 2008):

- *Paleo-weathering:* Weathering-leaching-evaporation (gypsum and calcium nodules).
- *Fissure type* (Fig. 5.6.2): Leaching-evaporation (near the ground surface, usually no more than several meters).
- *Quaternary sediments type:* (Leaching)-absorption (reduction)-evaporation (near the ground surface, usually no more than several meters).

The uranium resources are present in more than one northern Jordanian horizon. Pleistoces are the only horizons primarily composed of limestone and cherty fragments cemented with soil soft materials (Xoubi, 2008). The three other horizons are primarily composed of chalk and marl (Fig. 5.6.3).

Fig. 5.6.2 ***Fissure filling uranium (yellowish-green) deposits.***

Formation	Thickness (m)	Horizones	Uranium Shows (yellow staining)	Gamma Radiation	Lithology	Description
Pleistocene	At least 1 m	Horizon 1	Not seen	Reach up to 750 c/s		Limestone and cheert fragments cemented by soil (soft materials)
Upper most part of Muwaqqer Chalk Marl	3.35 m	Horizon 2	Moderate shows (yellow), its very finely distributed homogeneously in rok	Reach up to 1000 c/s		Marl, silt with pebbles of limestone
	5.0 m	Horizon 3 (concretion rich)	Highly shows of yellow staining of uranium	200–1000 c/s		Five concretion horizones cemented by fractured chalk that rich in uranium intercalated by red clay and gray marl
	2.0 m	Horizon 4	Low shows	100–150 c/s		Thick and massive beds of white chalk, low fractured with uranium shows but, very low

Fig. 5.6.3 *Uranium characteristics in North Jordan.*

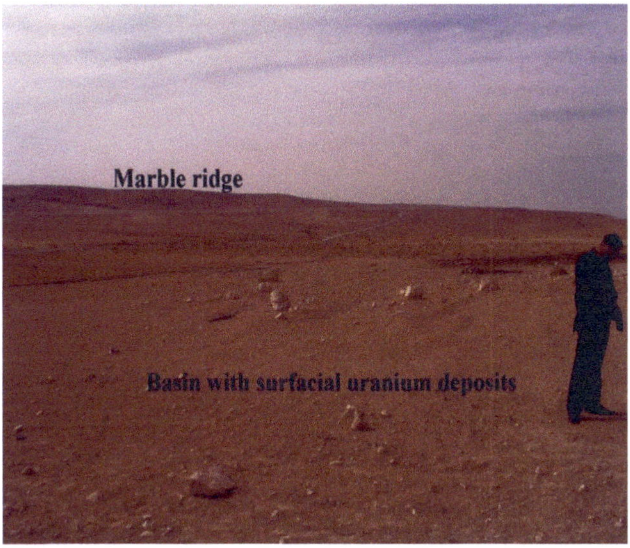

Fig. 5.6.4 ***Marble ridge surrounding basin with surfacial uranium deposits, Khan Az Zabeeb, Central Jordan.***

Khan Al Zabeeb area is located about 60 km south of Amman. It shows low relief terrain, mainly of sedimentary Maestrichtian age forming a basin. This basin is surrounded by low relief ridge formed of marble (Fig. 5.6.4), while the basin is filled with recent Pleistocene loose sediments.

The marble ridge is strongly tectonized, partly crushed, and dislocated by some low-angle faults (Fig. 5.6.5). It is altered; bleached with ferruginous highly altered zones. The secondary uranium mineralizations are present as filling fractures, patches, and staining the joint surfaces (Fig. 5.6.6). The marble is highly tectonized and highly altered. On the slope of this ridge, the secondary uranium mineralizations are noticed filling fractures and staining Pleistocene sediments. The basin is covered with loose Pleistocene-Recent sediments formed mainly of carbonate soil and contains silt, limestone fragments, and sands (Fig. 5.6.7). The radioactivity shows increase when going toward the basin and with depth. It ranges from 500 cps to more than 2000 cps in 30 cm depth (Fig. 5.6.8). Secondary uranium mineralizations were noticed staining some limestone fragments (Fig. 5.6.9) within this basin. It is noticed that some olive farms are located in this basin (Fig. 5.6.10). The area of this surfacial uranium–bearing basin is about 9.18 km^2, the average uranium content is about 946 ppm (Xoubi, 2008).

Fig. 5.6.5 ***Marble ridge with high secondary uranium mineralization, Khan Az Zabeeb, Central Jordan.***

Fig. 5.6.6 ***Strongly tectonized, partly crushed, and faulted marble ridge.***

Fig. 5.6.7 *Recent sediments with surfacial uranium deposits, Khan Az Zabeeb, Central Jordan.*

Fig. 5.6.8 *A shallow pit in recent sediments with surfacial uranium deposits, Khan Az Zabeeb, Central Jordan.*

Fig. 5.6.9 *Secondary uranium on limestone fragment, surfacial uranium, Khan Az Zabeeb.*

Fig. 5.6.10 *Olive farm in the basin of surfacial uranium, Khan Az Zabeeb.*

Phosphate deposits

The available information indicates that about 60% of Jordan area is covered with phosphate-bearing formations that are often exposed near surface or at various depths (Fig. 5.6.11). The uranium reserves in these phosphates are about 100,000 tons uranium according to the evaluation of IAEA in the Red Book. This reserve is not to be extracted through normal methods. This evaluation includes all phosphate occurrences in the country. The uranium content in these phosphate deposits ranges between 40 and 160 ppm.

It is noteworthy to mention that there is a radiometric disequilibrium in U^{238}in some phosphate deposits of El Hassa mine. This often indicates some washing for uranium in this locality which may be migrated to unknown sites. The phosphate deposits in Jordan include about 70000 tons of uranium, these amounts belong to the byproduct category, often from phosphoric acid. Eshdia deposits contain an average of 25–50 ppm of the uranium content, which constitutes most phosphate resources. Smaller deposits of Al-Hassa and Al-Abiad contain an average of 60–80 ppm of uranium.

Bituminous shale (oil shale)

This shale is present in Jordan at several locations such as El-Lajjun, Al Soltany, and Gurf Aldarweesh. These beds of oil shale are located at the base of chalk beds that are overlain the phosphate beds. The reserves of bituminous Shale that can be exploited by open pits mining are estimated as 40 billion tons. This amount includes 4 billion tons oil, in addition to other huge amounts at deeper levels. In the west of central Jordan, about 110 km south of Amman, and a midway point on the highway between Karak and Qatrana, is located the oil shale deposit at El-Lajjun. The deposit is 10 km long and 2–2.5 km in width. Situated on a trendy northsouth graben, this deposit is the shallowest known oil shale deposit, providing favorable conditions for open pit mining (Hamarneh, 1998). El-Lajjun oil stratigraphic units consist of limestone, marl, cherts, shales, and phosphates of Campanian–Maestrichtian to Holocene ages. The oil shale is 30 m in average thickness. The key structural characteristic is the El-Lajjun graben that controls a morphological depression bordered by faults striking of northsouth, generally from east and west. There is no information about the uranium contents of this bituminous shale. However, in some limited

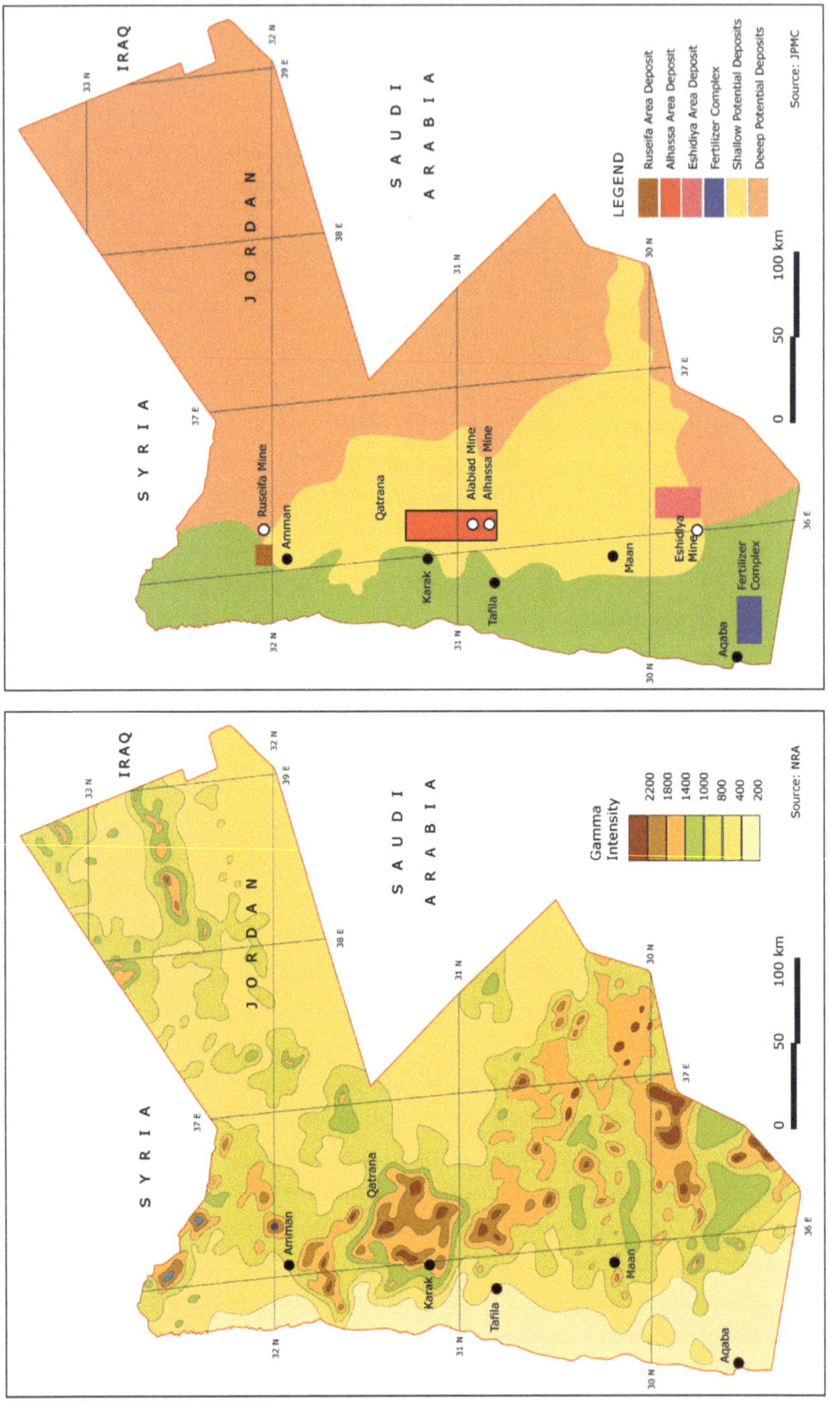

Fig. 5.6.11 ***Location of the phosphate mines in Jordan.***

analyses for samples of these shales, the uranium content ranges from 4 to 50 ppm (in Arabic). It is important to mention that the mode of occurrence of the oil shale adjacent to phosphorite deposits and its localization in remarkable graben structure raise its potentiality to host economic uranium deposit. This uranium deposit can be localized in deeper horizon where reducing conditions often prevailed.

CHAPTER 5.7

Saudi Arabia

Geological outlines

The Precambrian Arabian shield is the collection of exposed Precambrian rocks in the western part of the Arabian Peninsula. This shield represents the base for the sedimentary sequence cover in the northern and western regions of the Arabian Peninsula and is the continuation of the Nubian Shield in Egypt and Sudan on the western coast of the Red Sea. The two halves of the shield were separated during the Red Sea rifting during the Miocene Period (Fig. 5.7.1).

The contemporary state of Saudi Arabia covers the majority of the Arabian-Nubian shield, which consists mainly of basement rocks concentrated mainly in the western part of the Kingdom. The shield consists mainly of amphibolites, volcanic rocks, sedimentary and granite rocks with some basalt, which belong to the Tertiary volcanism in most cases (Duyverman, 1984). The basement rocks are covered from the east and the northwest following with the sedimentary rocks belonging to different eras, from the Palaeozoic to the Recent (Powers et al., 1966).

The greater Arabic basin, which occupies the broad Arab shelf and borders the Arabic shield, is one of the most important features in Saudi Arabia. It also contains several smaller basins. The clastics in this larger reservoir are identified in many locations on the margins of the basin as fine bands and scattered small exposures. The Late Permian carbonates in the Greater Arabian basin lie practically everywhere, consisting of sandstones, siltstones, shales, and thin argillaceous carbonate beds, including the Arabian Unayzah formation, Syria's Doubayat Group, the southeastern Turkish Hazro formation, the Iraqi Ga'arah formation, the Faraghan formation of

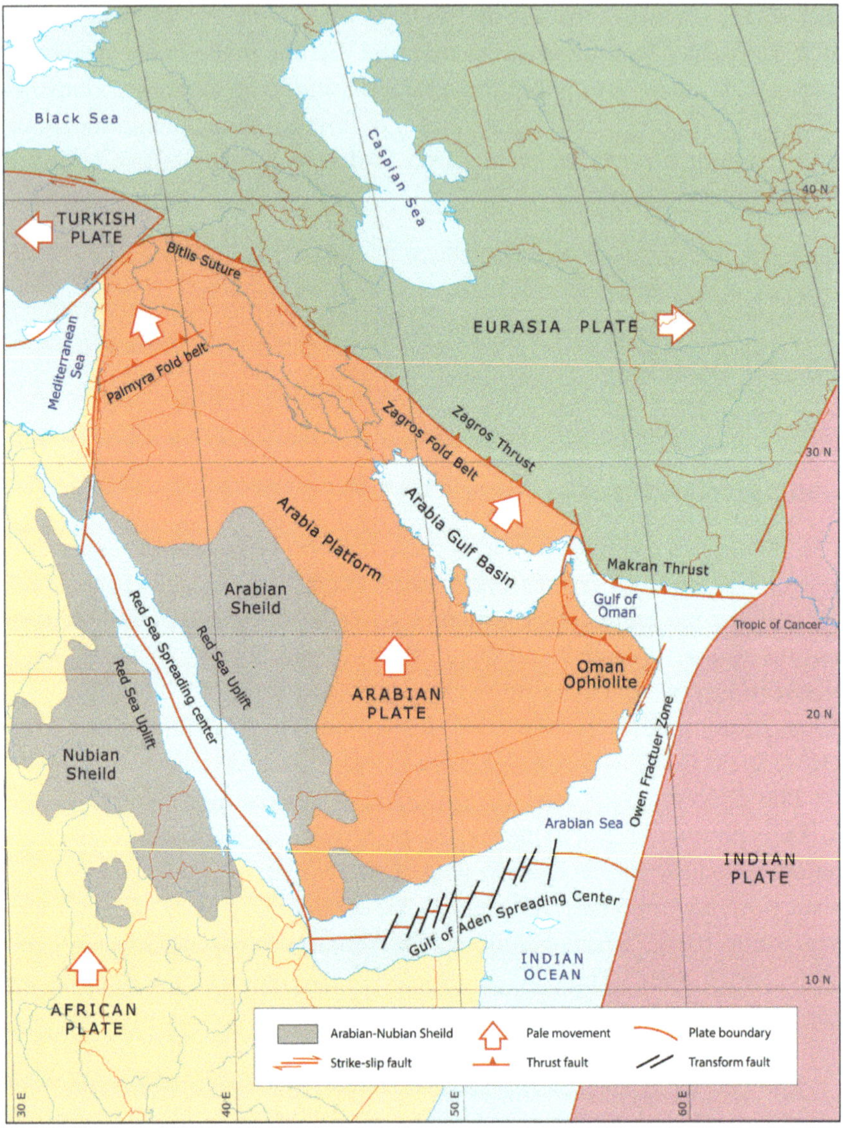

Fig. 5.7.1 ***Simplified map showing the main tectonic features in the Arabian-Nubian Shield (After Johnson, 1998).***

southwest Iran, and the Haushi Group of Oman (Al-Laboun, 1987). The good potential of this larger hydrocarbon basin is the source for its potential uranium resources.

Historical review on uranium exploration

There is little published information about uranium in Saudi Arabia, and most Lesbian radiological known radioactive anomalies belong to granite rocks (De Voto, 1982; Alfotawi et al., 1991). Despite the limited data available, it has been possible to collect some information on the evolution of uranium exploration in Saudi Arabia, and the following pages will identify some areas of special importance. Although there is no proof for the existence of economically viable uranium deposits in Saudi Arabia, according to the results of exploration work five areas have been identified as targets for uranium exploration. Those five areas are related to three geological environments: volcanic rocks in the El Roudah and Al Hanakiyah areas; in the sandstone of the Tabuk Basin, and in the recent wadi deposits in the El Holifah region; and in the layers of the lake sediment in the Sabkha and Aldumath areas. The layers of phosphate in the Toreif Basin contain some uranium, but its P_2O_5 contents are below economically viable levels according to current phosphate prices.

Uranium exploration began in Saudi Arabia since 1956. The knowledge of the geology of the Kingdom at the time was limited and funds for exploration were limited. Exploration began using a motorized radiation survey, which resulted in the discovery of some radioactive Lesbian but it was not considered important at the time. Airborne surveys in Saudi Arabia using scintillation devices and spectrometric instruments were carried out between 1961 and 1974, but those surveys were for the purpose of compiling magnetic data weather, with the speed of the aircraft and navigation data not allowing for the registration of radiation. In 1973, the sandstone of the Alwaged area was designated for exploration by the airborne radiometric survey to search for uranium. But the efforts made thereafter by the Newmont for Mines and BRGM firms failed to identify any uranium deposits in that region. In 1974, the systematic evaluation for uranium potentiality in the Kingdom and the necessary budgets were contracted with a Minatom. The implementation of this project was to be carried out in the following six stages:

- *Phase I:* Office work to collect data and assess the available information and determine appropriate geological environments.

- *Phases II and III:* Field regional exploration for those geological environments and identify the appropriate areas.
- *Phases IV and V:* Field exploration in areas selected in the previous two phases to identify target areas in a more precise manner.
- *Phase VI:* Analysis of the obtained data to identify areas of uranium deposits in the target areas.

This project has identified several important targets for the exploration of uranium: the hydrothermally altered volcanic rocks of the Proterozoic era of the Arabian Shield, an area of sandstone from the Palaeozoic era, and two others which represent concentrations close to the surface in the recent unconsolidated beds. The phosphate beds in the Turev Basin contain between 15 and 40 ppm uranium and rarely reach 200 ppm. It is worth mentioning that there are techniques that can be applied to extract uranium as a byproduct from phosphoric acid, but experience in many places shows that when uranium concentration and the current prices are low, it is difficult to extract uranium economically (Salman, 2005).

Uranium resources potentialities

The review of the available data related to uranium exploration in Saudi Arabia, as well as the regional vision in the Middle East area, indicates that uranium deposits are linked to several different geological environments in Saudi Arabia. The most important of these are the volcanic rocks and tuffs that cut in the younger Pan-African granite within some fractures and fault zones of the Najd system. This is in addition to the presence of recent calcrete uranium deposits.

According to the main criteria of the regional geological units, as well as the prevailing geological structures in Saudi Arabia, the high uranium resources potentialities could be associated with two main types: the Pan-African younger granites, especially those associated with some acidic volcanic rocks and tuffs, and the Palaeozoic sedimentary rocks sequence. Therefore, these provinces are recommended to host uranium resources.

Pan-African younger granites

Numerous exposures of these granites are present in the northeast, the southwest and northwest of the Arabian Shield (Riddler et al., 1986). These younger granites represent a source for uranium deposits. Studies on 17 younger granite intrusions in the northwest of Saudi Arabia have shown that those rocks contain radioactive anomalies (Stuckless et al., 1983).

Further study on some granite intrusions in the Madian Area has shown the presence of high uranium and thorium contents, proving the presence of uranophane in the site of this radioactive anomaly (Alfotawi et al., 1991). It is therefore evident that the granite rocks of the Arabian Shield in Saudi Arabia are considered one of the potential sources for uranium resources. It is important to focus the exploration efforts on the per-aluminous, double mica rich in Mo, F, and element granites which belong to the Late to post-orogenic igneous activities. The contact zones and peripheries of these intrusions represent good target for exploration. Egyptian experience in exploration has proven that most uranium mineralization is concentrated in areas of the younger granite plutons edges (Salman et al., 1990). In these younger granites, economic deposits of uranium ores can be found, but with small- to medium-sized type. It is also believed that attention should be paid during exploration to acidic and alkali volcanic rocks that extrude from the younger granites in the Arabian Shield. It is also important to pay attention to sites of intersection of the Najd fault system that strikes in the direction of the northwest–southeast with those faults following eastnorth, eastwest, and southwest direction. The intersection zones are a suitable environment for the presence of local uranium deposits if geochemical conditions are suitable.

The Asfar Thwelil uranium occurrence

The uranium mineralization is present in the form of thin veinlets of uraninite (Fig. 5.7.2) and pitchblende injecting the rhyolite rocks of Jebel Asfar Thwelil in the northern Arabian Shield, Saudi Arabia. These primary uranium minerals are associated with secondary uranophane (Fig. 5.7.3) uranium mineral, fluorite, thorium, and REEs minerals. These uranium veinlets are present only in the domal part of the rhyolite, which shows flow texture nearby the caldera, but never detected until recently in the caldera itself. This rhyolite is characterized by a high uranium background. This type of uranium mineralization is recorded in several parts in the world and is known as volcanogenic uranium deposits (Goodell and Waters, 1981). It is noteworthy that the Asfar Thwelil area is located in the northern part of the Arabian shield and contains a rhyolite dome on the edge of a volcanic caldera, which is filled with ignimbrites that had been partially reworked. These rocks belong to the Shammar Group that is the youngest acidic volcanic unit in the Precambrian Arabian shield. The area contains three different types of mineralization: (1) Mn- and Fe-oxide veins in the clastic

Fig. 5.7.2 ***Uraninite veinlets injecting the rhyolite, Jebel Asfar Thwelil, north Arabian Shield, Saudi Arabia.***

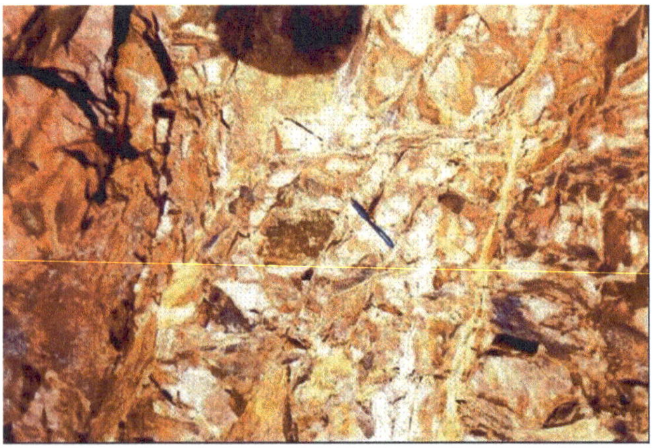

Fig. 5.7.3 ***Uranophane veinlets and patches injecting the rhyolite, Jebel Asfar Thwelil, north Arabian Shield, Saudi Arabia.***

rocks on the edge of the caldera, (2) disseminated mineralization of precious and base metals (plus Mn) in the caldera pyroclastics, and (3) veinlets of uranium cutting the rhyolite dome (Ben Atiq, M.A, 2005).

Palaeozoic sedimentary rocks

The Palaeozoic sedimentary rocks are exposed in the northwestern Saudi Arabia in general (Figs. 5.7.4 and 5.7.5). This stratigraphic sequence and

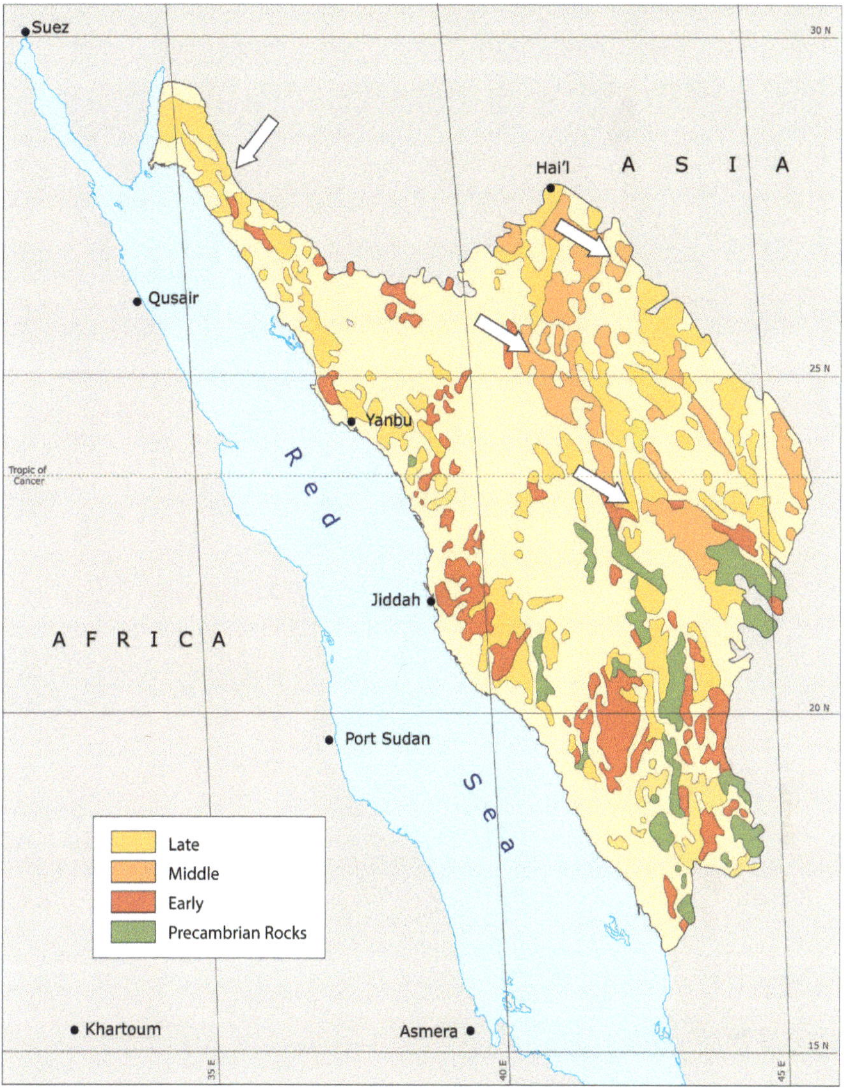

Fig. 5.7.4 ***Map showing the younger granites distribution with high uranium potentiality in Saudi Arabia.***

mineral components of these rocks must be studied to identify the suitable criteria for the localization of uranium ores, especially the sandstone uranium type.

It is noteworthy to mention that secondary uranium mineralizations were recorded in the Abu Znimah area in the western Sinai of Egypt in

Fig. 5.7.5 ***Simplified map for a part of Saudi Arabia showing the high uranium potentiality areas.***

some the Lower Carboniferous beds (Afifi, S., 1994). The existence of some high radioactive anomalies in the Wadi Araba area, north East Desert of Egypt, in similar rocks has also been revealed (Salman and El-Aassy, 1983). Therefore, it is believed that the Palaeozoic rocks in northwestern Saudi Arabia can be considered as a good target for the exploration of uranium resources, where there is a good potentiality for the presence of sandstone uranium type. It should be noted that this type of uranium ore can be mined at its location (in-site leaching). Thus, the extraction of uranium there will be inexpensive and will likely contain elements of other secondary precious elements that can be extracted with uranium as a byproduct, also helping to reduce costs.

Sabkhas uranium deposits

It is important to pay more concern to the sabkhas areas, particularly those located in closed or semi-closed basins and near to younger granites or acidic and alkaline volcanics that have high uranium content. The uranium in this kind of recent deposits can be easily and cheaply extracted (Salman, 2005).

Phosphate deposits

North Saudi Arabia's phosphate deposits are organic Mesozoic age sediments that are gently folded and cover Paleozoic sediments. Al-Jalamid's. The unweathered phosphate ore is dolomitic and not found on the surface. The de-dolomitization of compact calcareous ore forms the friable surface layer of the deposit and the gradual calcite solution of the calcareous form ore. Laterite weathering in the phosphate deposits west of al-Jalamid breaks down the uranium-contained apatite to form aluminum phosphate minerals that have been deposited as secondary yellow uranium minerals in and around the phosphate. It was not observed in al-Jalamid, but the gamma-ray log indicates that in the near-surface area the weathered uranium-phosphate ratio increased in the field. The more porous, friable ore provides better access to uranium gases and greater surface exposure to the device, which artificially raises the apparent uranium/phosphate ratio (Wynn, J. 1996). The phosphate deposits in the northern area of Saudi Arabia represent an extension of the Jordon phosphates that include uranium resources. This fact often supports the presence of uranium resources in or near phosphate deposits in Saudi Arabia.

CHAPTER 5.8

Sudan

Geological outlines

The exposed rocks in the Sudan are represented by Precambrian basement complex, Mesozoic, and Tertiary sedimentary rocks. The basement complex is composed of granodioritic quartzo-feldspathic gneisses, migmatites, amphibolites-grade metasediments, gabbroic bodies, and syntectonic granites. These rocks have been intruded by a suite of anorogenic alkali granites and syenites that represent the last major magmatic event in the area. The surrounding unmetamorphosed sedimentary units include the Mesozoic. The Nawa formation is composed of mudstone and minor limestone, the Cretaceous Nubian Sandstone formation, and Tertiary-Quaternary unconsolidated sediments (Sadig et al., 1988).

Uranium resources

Limited information about uranium resources in the Sudan is available. Also, there are no country reports for the Sudan in the uranium books of the IAEA Uranium Group meeting. One radiometric anomaly has been located in Uro, Nuba Mountains, following a zone of tectonic breccias. Maximum levels are as follows:

U = 2220 ppm P_2O_5 = 26.1% Al_2O_3 = 35.4%
V = 3080 ppm Ba = 14000 ppm F = 7900 ppm

Sporadic highs in Cr, Cu, and Zn occur in aluminous phosphate (crandalite). The tectonic breccias are confined to a long, steeply westward-dipping fracture zone that cuts the graphitic slates. The width of the zone varies from 7 to 40 m and it is 1.5 km long. The rocks occupying the zone are made up of angular fragments of slate, quartzite, and marble (Sadig et al., 1988).

Other radioactive anomalies have been reported in the Miri locality, located in the Kadugli area in the southern part of the Kordofan Province in the Nuba Mountains. The mineralized veins are hosted in granite rocks, where thorium dominated, but do contain anomalous amount of uranium. The uranium content ranges from 13 to 300 ppm, while thorium content ranges~ from 5 to 769 ppm. The mineralized veins are rich in Zr = 113 81, Nb = 40 17, Y = 1160, Ce = 5168 and La = 1092 ppm

(Parslow et al., 1992). The plans for uranium prospecting in Southern Sudan are being promoted by the Brinkley Mining-Southern Sudan. Southern Sudan's geology is made up of a Precambrian Basement, which is mineralized by its various structural elements. Recent research work has identified five to six times of a radioactive readings. The key factor responsible for uranium mineralization in this region is assumed to be late-stage Pegmatite intrusions.

Potential uranium resources

Considering the available information on uranium and the regional geological and tectonic aspects, some geological environments can be considered as targets for potential uranium provinces in the Sudan. These are intra-cratonic rift basins and granitoid rocks.

Pan-African granites

The granite rocks in the Sudan are represented by bathonthic granites and younger granites. The batholithic granites are of calc-alkaline type. The younger granites are post tectonic, epizonal alkali, MgO-poor, and highly silica saturated (Ghuma and Rogers, 1978). The granitites, especially the younger ones, can be considered as potential uranium provinces. It is worth mentioning that some granite and syenites rocks in the Nuba Mountains show a high content of uranium and thorium (Sadig et al., 1988 and Parslow et al., 1992). Therefore, the post-orogenic granites of the Sudan could be considered as target provinces with good potentiality for vein and disseminated-type uranium deposits.

Intra-cratonic rift basins

These rift systems were developed during Cretaceous tension tectonics in the southern Sudan. In these basins, a thick sequence of non-marine rocks accumulated. These rocks are represented by fluvio-Iacustrine arkosic sandstones. Moreover, the associated shales and claystones deposited in the anoxic environment contain organic matter in the range of 5%–20%. These rifts are represented by Muglad (Abu Gabra), Meult, and Blue Nile basins (Fig. 5.8.1). The Abu Gabra basin joins with the Central and West African rift system through the south of the Chad rift in the central part of the continent. Most of the discovered oil fields in southern Sudan are concentrated in Abu Gabra basin (Nagati, 1986).

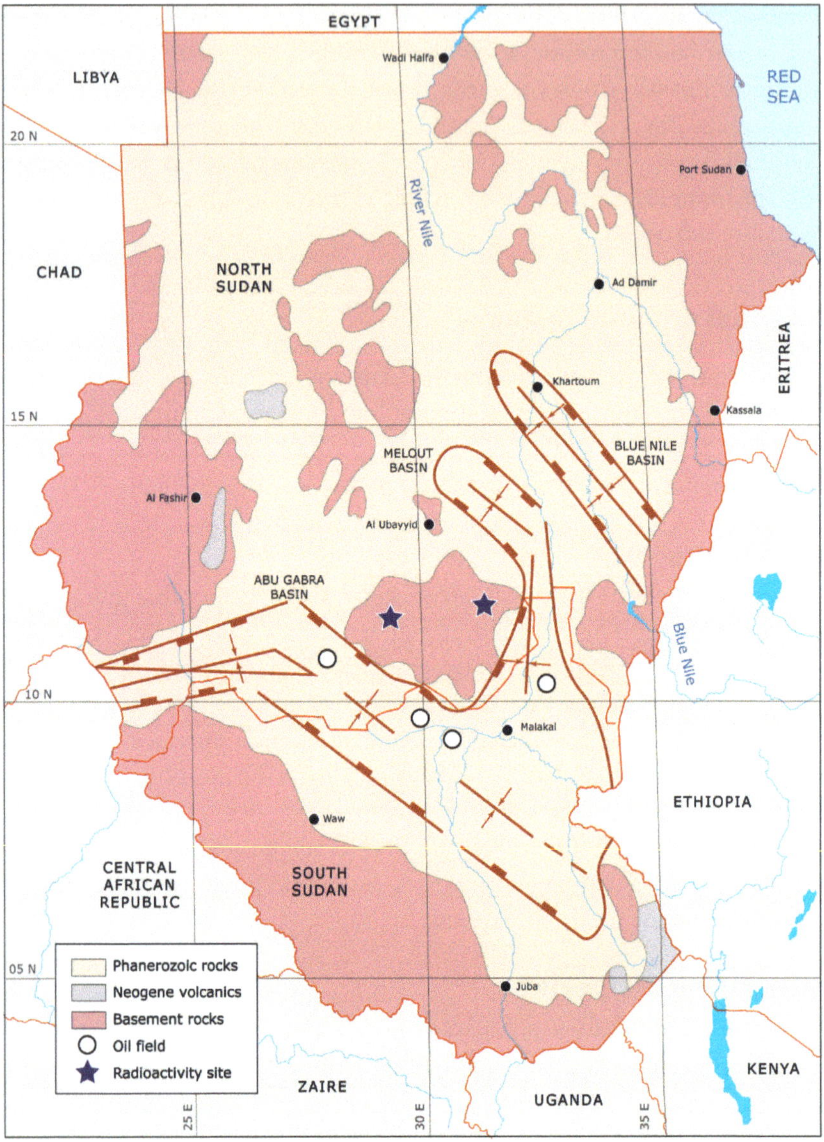

Fig. 5.8.1 ***Geological map of Sudan with some rift basins showing the location of radioactive anomalies and potential Abu Gabra rift basin.***

These rifts, especially Abu Gabra, can be considered as a potential uranium province where it has most of the recognition criteria for the identification of sandstone-type uranium deposits. In addition to the presence of thick continental sandstone associated with organic matter–rich shale beds, the presence of oil can act as a reluctant factor. Also, these basins are surrounded by a large terrain of basement rocks that include some uranium-rich varieties that act as a good source for uranium.

CHAPTER 5.9

Syria

Geological outlines

The surface geology of Syria varies from one region to another according to the tectonic framework of the country (Fig. 5.9.1). In the eastern

Fig. 5.9.1 ***Space image showing the main tectonic features of Syria.***

area, the Neogene sedimentary units are exposed. The Upper Miocene gypsum, limestone, marls, clays, and sandstones prevail at the border zone with Iraq. In the area of Euphrates River, continental conglomerates of Pliocene age and Upper Miocene sediments are exposed. Moreover, some Quaternary basalt is outcropped at the northern environs of this eastern sector. The outcropping of this younger sedimentary sequence is often controlled by the Great Euphrates graben structure which follows a NW-SE trend.

The central part of Syria includes rock exposures that mainly belong to the Upper Cretaceous sedimentary sequence that mainly follows an ENE-WSW extension trend. These sedimentary rocks are mainly represented by dolomites, limestones, clays, chalk, and organogenous detrital limestone intercalated with clay. The main Upper Cretaceous phosphate deposits of Syria are concentrated within this area. These Upper Cretaceous units are flanked from the north and south by Eocene and Miocene sedimentary units. It is important to note that some phosphate deposits are included within the Eocene units.

The exposures of the sedimentary units in the central part of Syria are often controlled by the main tectonic framework of Syria. In this area, the Palmyrides uplift follows the ENE-WSW direction. The main Upper Cretaceous phosphate deposits are included around this promising structure. Moreover, the Eocene phosphate deposits are located to the south of this structure at the Rutbah uplift (Figs. 5.9.2 and 5.9.3).

In terms of tectonics, throughout northwestern Syria, there are few basement exposures (Ponikarov, 1966 and Sawaf et al., 1993). The basement is 8.5 km deep in the Rutbah uplift, about 9 km deep in the Euphrates depression, and between 5.5 and 6.5 km deep north of the Euphrates. Therefore, the extensive thickness of pre-Mesozoic rock is documented. The structure of the Euphrates graben that reveals the thick-skinned tectonic design of this area describes profoundly penetrating weaknesses. The Late Proterozoic accretion of the northern Arabian platform with the Euphrates fault system can be considered as a suture or shear zone. This can be confirmed by gravity observations and measurements, which indicate the higher density crust in northeast of the Euphrates pattern—a typical feature for the suture areas. This is the theory behind the Late Proterozoic accretion Euphrates' fault network. This supports theories of a network of vulnerable areas underneath the northern Arabia platform, which is inherited from Late Proterozoic/Early Cambrian accretion, which tend to influence the tectonics or regional areas.

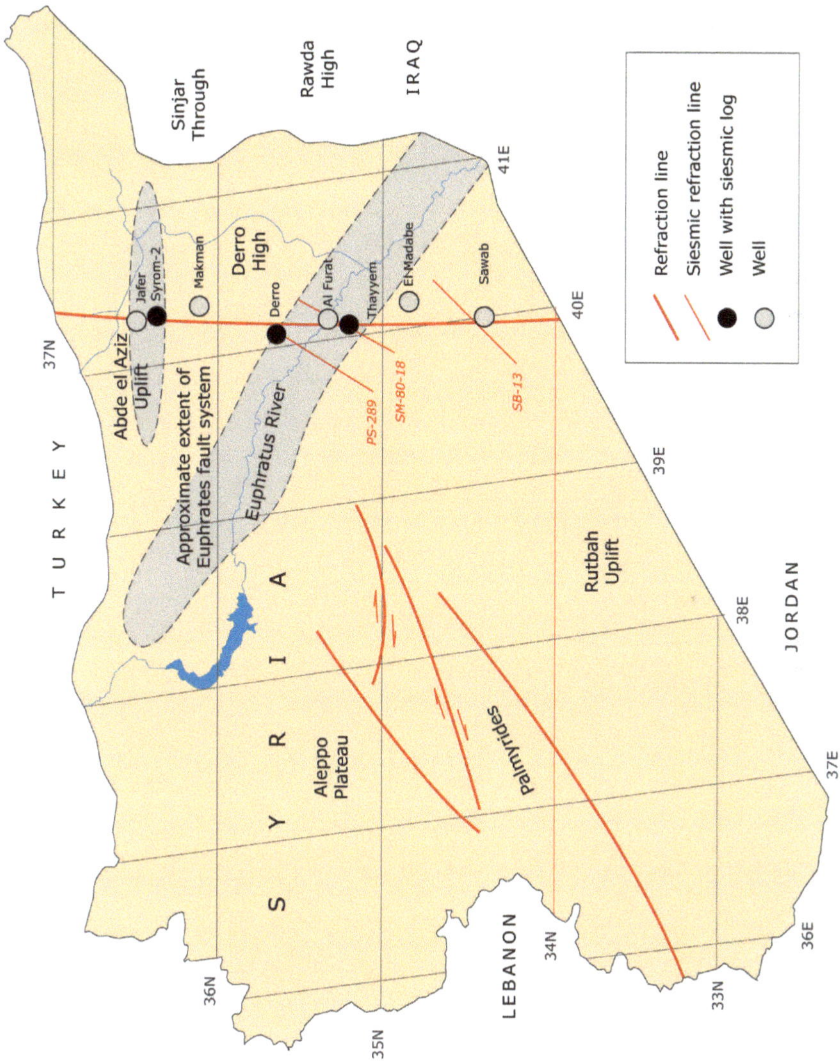

Fig. 5.9.2 ***Tectonic map of Syria.***

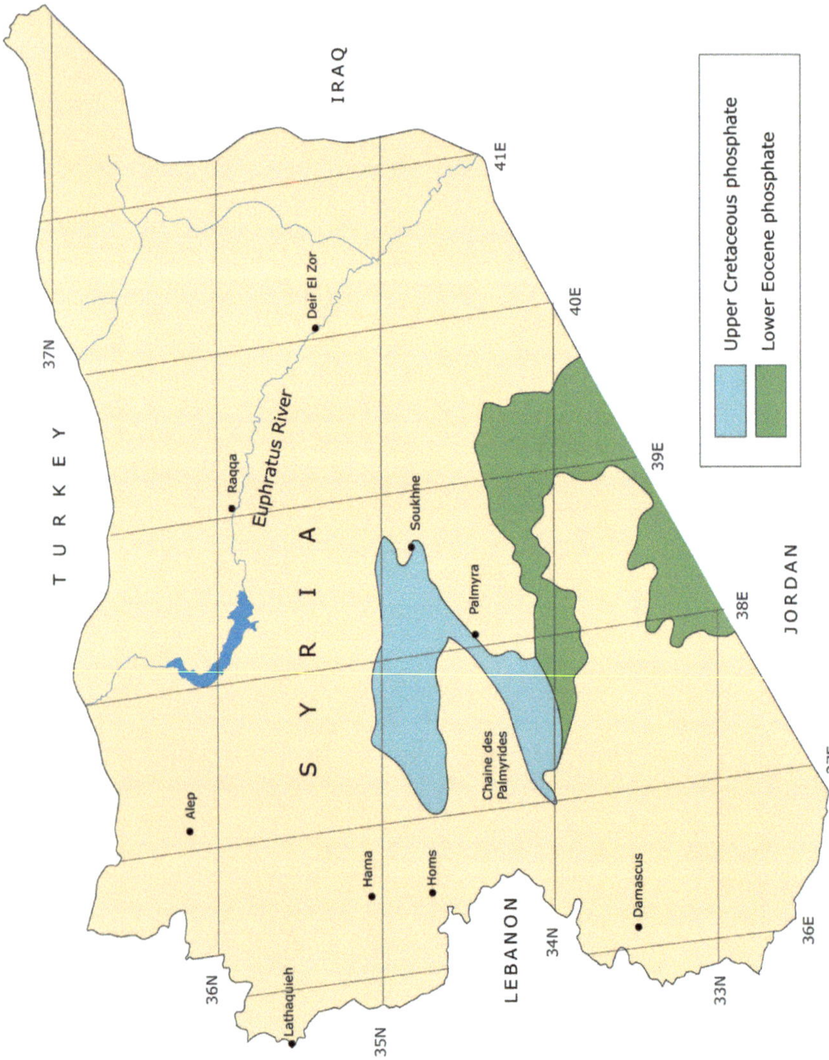

Fig. 5.9.3 ***Map showing phosphate exposures.***

The folding Palmyride belt is a 400 by 100 km northeast-trending belt located in the central Syria, in the northern Arab plains, and bordered by the Aleppo plateau in the north and the Rutbah uplift in the south (Fig. 5.9.1). The Palmyrides belt obliquely intersects (at about 45°) in western Syria and eastern Lebanon, the roughly north-trending Dead Sea transform fault system.

Second, proof of strike-slip displacement reported in both the field and academic literature and recognized by new focal mechanism solutions for two of the recent Palmyrides earthquakes suggested that some of the "missing" displacement in central and northern Syria could be distributed in a strike-slip motion oblique to the relative northward convergence of the Arabian plate on the Eurasian plate. Alternatively, previous slip estimates in the north section of the Dead Sea transform fault system may be just minimum estimates. A final possibility is that the Dead Sea transform fault was active in northwestern Syria for just the past 5–6 million years or so, meaning that it was either not present or shifted only before Pliocene (Chaimov et al., 1990 and Brew et al., 2000).

Uranium exploration in Syria

Uranium exploration in Syria began in 1960 and was concentrated on limited exposures of sedimentary rocks. Exploration works were completed after preparing a geological map of Syria. In the seventies, limited uranium exploration works started especially in the phosphate areas in the Tadmor Plateau.

The Syrian Atomic Energy Authority began an extensive exploration for nuclear materials, especially uranium, in 1982, following a workshop that included international experts and planners. This plan defined the following geological environment for uranium exploration:

1. Areas of phosphate exposure in the Tadmor Plateau.
2. The Quaternary Neogene rocks exposures near phosphate deposits.
3. Sabkha areas near phosphate exposures.
4. Igneous rocks in the NW part of Syria and volcanic rocks in the south and middle part of the country.
5. Clastic rocks and rocks that contain organic materials at the base of the Cretaceous sequence.

Exploration for uranium in Syria has applied various methods including car-borne radiometric survey, foot prospection, and emanation methods for radon gas detection. The airborne survey led to the discovery of two principal areas with highly radioactive anomalies in the El-Awaid area (Wadi El-Rakheem, northern Tadmor Plateau) and Thelithwat El-Gannah El-Segary in El Badiah. The report indicated that these radioactive anomalies are associated with phosphate exposures in these two areas. Moreover, the car-borne survey and field check indicated that the recorded radioactive anomalies are associated and concordant with phosphate beds exposures (Fig. 5.9.3).

The detailed fieldwork indicates the presence of some high radiometric readings and presence of weak uranium concentrations as crusts on the sabkhas at the peripheries of the El-Dow Basin of the Tedmor Chain. These crusts include secondary uranium mineralization as a result of washing and the movement of uranium through some hydrogeological systems in partially closed basins such as the El-Dow Basin and Sabkhet El-Mooh. The use of radon gas measurements in uranium exploration shows the presence of some high readings of radon gas in various areas of Syria. In the El-Ghab Plain, the high value of radon gas is associated with some directions of the important faults and tectonic fractures as the northern continuation of the El-Mashriq faults (Dead Sea Faults). In the area of Kherbet Hannorah, high radiometric anomalies are associated with secondary uranium mineralization at the contact zone between the marine Paleogene and the continental Neogene sedimentary rocks. Moreover, at the El-Teyas area, which is located on the Tadmor-Homs highways, some high concentration of α-tracks are recorded also along the contact between the Paleogene and Neogene units.

Potential uranium resources

The following can be determined from reviewing the available geologic, tectonic and radiometric data on Syria:

Phosphate deposits

Phosphate deposits represent the main source for uranium in Syria which has been mined for many years. Some of these phosphate deposits are used in phosphoric acid production in Syria. The average content of uranium in the main Cambanian phosphate deposits range between 60 and 100 ppm, while the Paleogene phosphate deposits contain about 37 ppm uranium as an average content.

It is clear from the above that there is low uranium content in the phosphate deposits in Syria, thus the direct extraction of uranium from them is not economically viable, but uranium can be extracted as a byproduct during phosphoric acid production. It is clear from reviewing some available information that Syria constructed a small unit for producing yellow cake from phosphoric acid. This unit should be developed to increase its capacity for treatment and purification of phosphoric acid for producing more yellow cakes.

It is clear from the studies performed on uranium distribution in Syrian phosphate deposits that some high uranium concentrations are present in fractures and joints. The uranium content sometimes reaches 3000 ppm. It is therefore important to study in detail the structures of the system in the Tadmor Chain and el-Dow Basin. It is also essential to perform geophysical studies for the delineation of the phosphate deposits at depth especially at the oxidation reduction boundary where the primary uranium minerals can be hosted.

Eastern tectonic zone

The application of radon gas measurements shows remarkable high values in the el-Ghab Plain and Kherbet Hannorah. The el-Ghab Plain represents a strongly tectonized zone and it is considered a continuation of the Eastern Fault (Dead Sea Rift). This area can be considered of prime importance due to the high readings of radon gas relative to other areas. An extensive uranium exploration is recommended in this area where detailed geological and radiometrical scans can be done. Also, a detailed geophysical ground survey is recommended which can detect an ore body in deep-seated zones where the reducing environment prevails. In reference to Kherbet Hannorah, which lies at the contact zone between Paleogene marine sediments and Neogene continental sediments, some high radon gas anomalies are recoded. These anomalies are associated with secondary uranium mineralization at several trenches. This site and the associated features are very interesting and further uranium exploration should occur here where detailed geologic, structural, and radiometric mapping should be constructed. This step should be followed with detailed geophysical studies for the contact zone on various depths, and then exploration core or percussion drilling to reach the oxidizing–reducing zone where an economic uranium ore deposit could be detected.

Recent uranium deposits

The presence of secondary uranium mineralization associated with phosphate deposits along some fractures and the presence of high radon gas anomalies along the eastern great fault zone often suggest the presence of recent uranium deposits in nearby areas. Moreover, the areas near the Tertiary volcanic activities sites can represent promising targets for uranium exploration. It is important to pay great attention to the exploration of uranium in sabkha especially in the close or semi-closed basin in the El-Dow Basin near areas of phosphate deposits. It is possible to perform more detailed works on the basins that gave preliminary encouraging results.

It is worth mentioning that the discovery of surface economically viable uranium deposits in Jordan in several sites gives a good potentiality for discovering similar uranium deposits in Syria. Moreover, the presence of huge phosphate deposits in Syria, the recording of several volcanic activity cycles, the extension of the Dead Sea Rift, and the associated tectonic and hydrothermal activity mean a high probability of finding economic uranium deposits. Also, the tectonic framework of Syria, which includes important tectonic features such as the Great Euphrates graben structure, the Palmyrides fault zone, and the Zagros fold belt and the associated regional thrust fault (Fig. 5.9.4) at the northern border of Syria, increases the probability of hosting economic uranium deposits. Middle East's regional tectonic map is shown in Fig. 5.9.5.

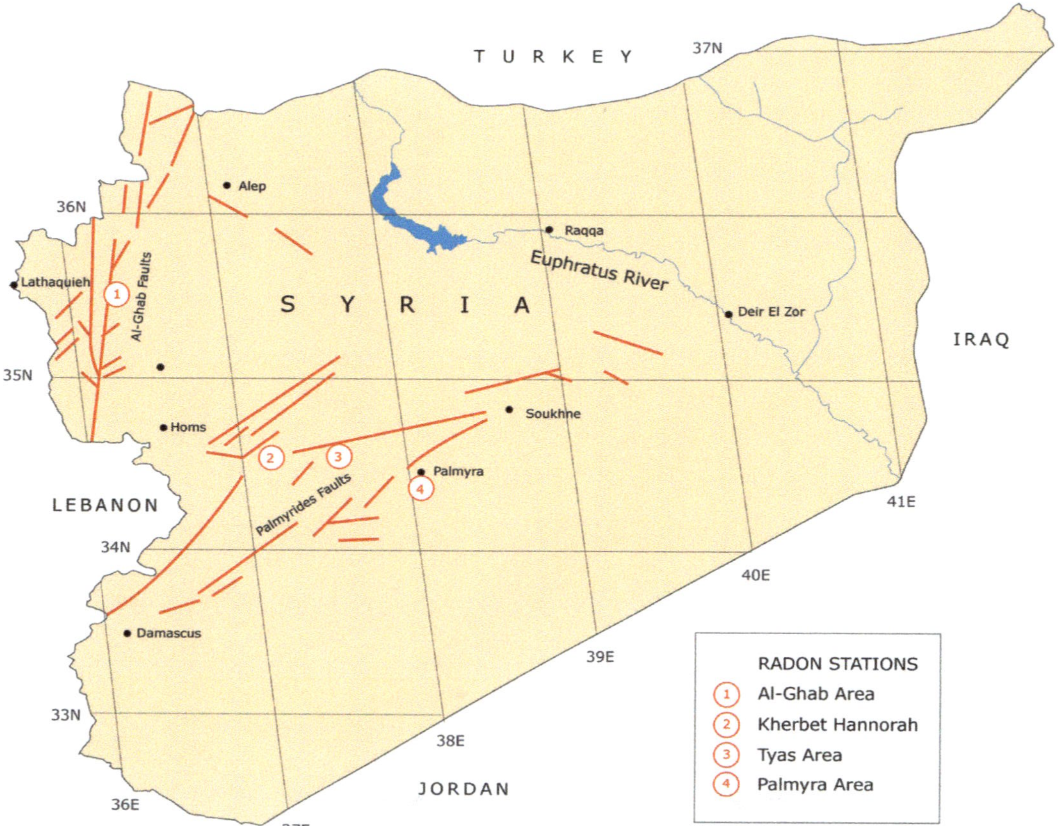

Fig. 5.9.4 ***Map showing sites of radon gas measurements in Syria.***

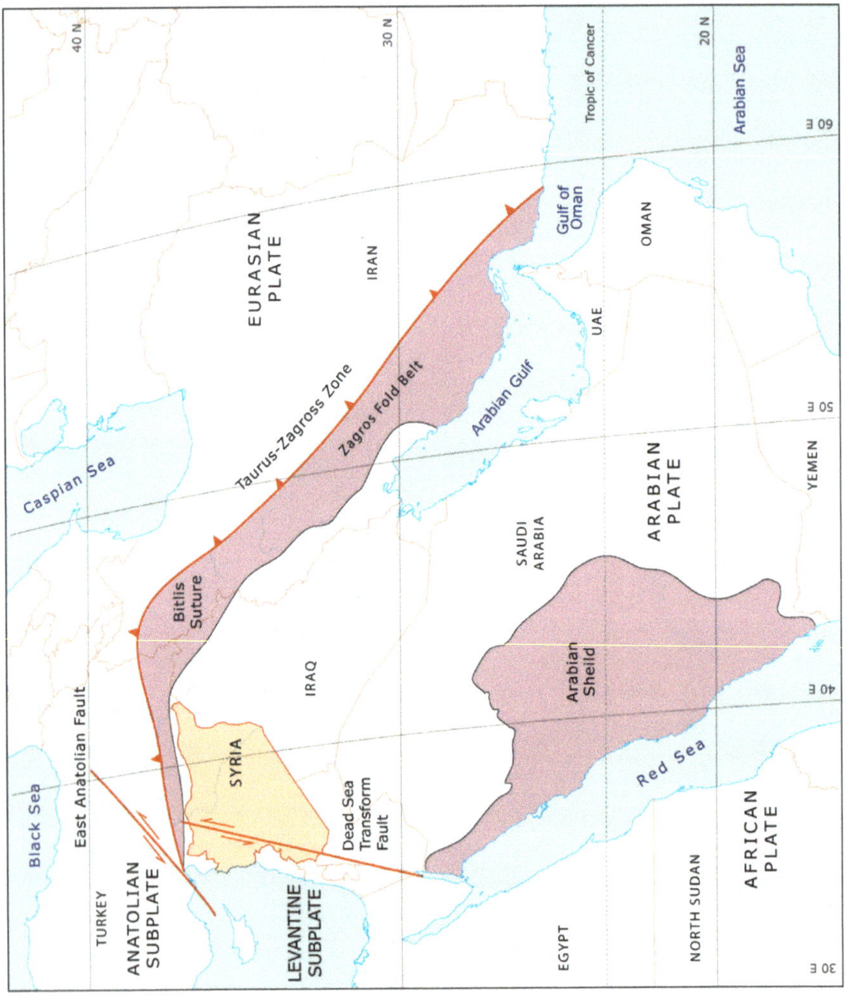

Fig. 5.9.5 ***Middle East's regional tectonic map.***

CHAPTER 5.10

Turkey

Geological outlines

Turkey is located in Anatolia and Southeastern Europe. The portion of Turkey west of the Bosphorus is geographically part of Europe, with the eastern portion of Anatolia considered part of Southwestern Asia. The country borders Bulgaria, Georgia, Greece, and Syria, and has coastlines along the Black Sea, the Aegean Sea, and the Mediterranean Sea. The country's geographical coordinates are 39°00′N, 35°00′E, covering a land area of 780,580 square kilometers. Of which 770,760 square kilometers are land, and 9820 square kilometers are water. Fig. 5.10.1 shows the regions of Turkey.

Geologically, Turkey is part of the vast Alpine belt from the Atlantic Ocean to the Himalayas. This belt was formed during the Tertiary Period (approximately 65 million to 1.6 million BCE), when continental Arab, African, and Indian plates began to collide with the Eurasian plate, and the sedimentary layers laid down by the prehistoric Tethyan Sea buckled, folded, and contorted. This mountain belt was intensively folded and raised, followed by extreme volcanic activity and intrusions of igneous rock material, followed by extensive faulting during the Quaternary Period, which started about 1.6. million BCE. This process of folding and faulting continues today as the Turkish and Aegean plates continue to collide, both heading south and southwest. It makes Turkey one of the most active areas of the world for earthquakes and volcanoes.

Turkey has Precambrian rock outcrops (Fig. 5.10.2) over 540 million years old (Bozkurt et al., 2000). Turkey can be thought of as a collage of different lands from the ancient continent and oceanic lithosphere, along with younger igneous, volcanic, and sedimentary rocks. During the Mesozoic period there was a large ocean (which existed from about 250 to 65 million years ago) known as the Tethys Ocean between the supercontinents of Gondwana and Laurasia, with an oceanic lithosphere flooring (which lies to the south and north, respectively. This large oceanic plate was consumed in the subduction zone. The subduction trenches were folded, faulted, and tectonically combined with large blocks of crystalline basement rocks of the oceanic lithosphere, which were deposited within in the ancient prehistoric Tethys Ocean. These blocks form an extremely complex mélange of rocks which mainly consists of serpentines, basalt, dolerite, and

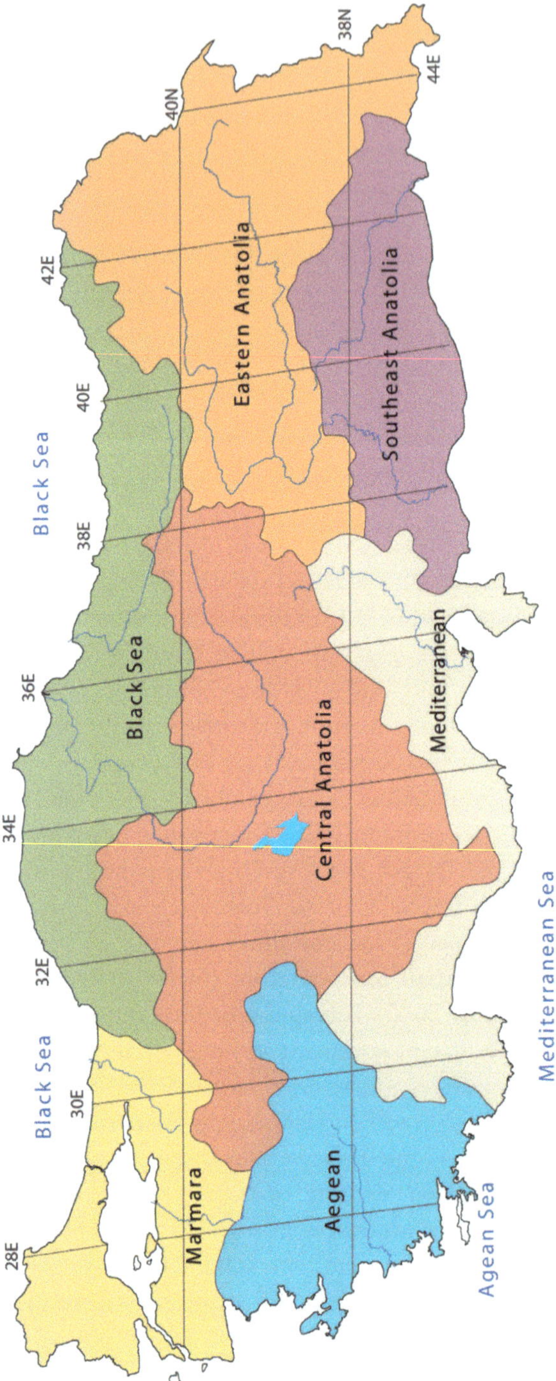

Fig. 5.10.1 ***The main regions of Turkey.***

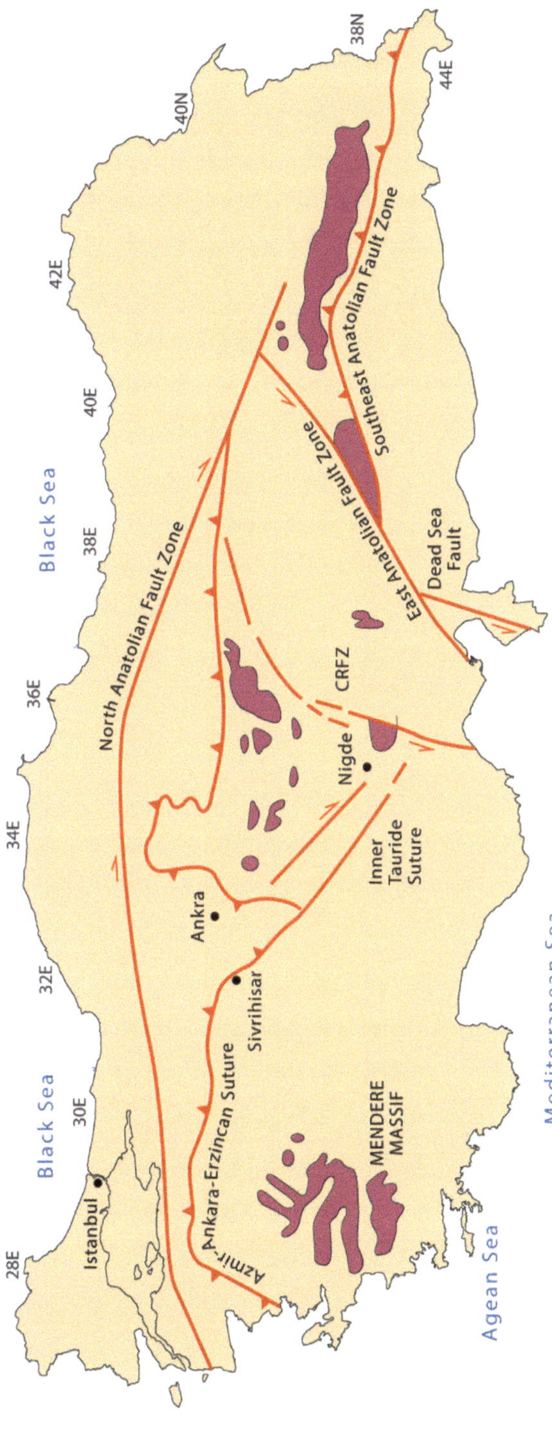

Fig. 5.10.2 ***Simplified geological map of Turkey showing the location of some of the places in the Central Anatolian Crystalline Complex (CACC) and a region near the town of Sivrihisar.***

chert. Ophiolites (see ophiolite) volcanic arcs and backarc basins formed and settled on Eurasia as they collided with micro-continents (literally, relatively small plates of continental lithosphere; e.g., Ustaomer and Robertson, 1997). Such microcontinents were brought further south from the continent of Gondwanan. Thus, Turkey is composed of numerous prehistorical microcontinents. During the Cenozoic Period (Tertiary around 65–1.6 million years ago) the folding, faulting, and uplifting, followed by volcanic activity and intrusion of igneous rocks is attributed to a significant continental collision between the larger Arabian and Eurasian plates (Robertson and Dixon, 1984).

The terrain of Turkey is structurally complex. A central massif is wedged between two folded mountain ranges that converge in the east, consisting of uplifted blocks and folded troughs, surrounded by recent deposits and giving the appearance of a rugged plateau. True lowlands are limited to the Ergene Plain in Thrace, extending along rivers that flow into the Aegean Sea or the Marmara Sea, and a few small coastal strips along the Black Sea and Mediterranean Sea coast.

Uranium exploration in Turkey

Turkey has concentrated its uranium exploration efforts to eliminate the possibility of becoming dependent on external uranium resources for the three nuclear power stations it plans to build in 2015 and therefore it is not adversely affected by price hikes. As the unit cost of mining operations for the 9120-ton uranium reserves that were previously discovered in Turkey is higher than global prices, the Mineral Research and Exploration Institute (MTA) decided to conduct new exploration surveys to discover more economical uranium deposits. During the last two years, the MTA has found new uranium deposits in Kırşehir, Nevşehir, and Ankara.

After developing the Radioactive Anomaly Map of Turkey, the MTA estimated that Turkey will not be dependent on external uranium resources during the initial operation of its planned nuclear power stations. Turkey will also establish a facility for uranium enrichment for its future nuclear power stations. After the Nuclear Technology and Energy Development Project, prepared by the Turkish Atomic Energy Agency (TAEK), was presented to the Prime Minister, Turkey has tried to decide where to establish the nuclear power stations and how to provide the raw materials for them. The TAEK report suggested eight different sites for the construction of nuclear power stations, and it is almost certain that the first station will be established in Yalıncıgerme, Sinop. In determining suitable sites for the nuclear power stations more than 40 parameters, including the provision of cooling water, convenience of transportation, lack of civilian aviation activity, the social environment, earthquake or tsunami risks, and international security were taken into consideration. Despite the reactions from environmental organizations, the nuclear power station tender was opened in 2008 according to the government's plan.

Turkish Atomic Energy Authority (TAEK) and the MTA have made important progress with respect to uranium to be used in the nuclear power stations. With joint projects, these institutions have managed to identify Turkey's uranium potential. With the air survey, the MTA discovered important uranium deposits in the Eastern Black Sea region (Rize and Artvin), Kırşehir, Nevşehir, and Ankara. In previous surveys, Turkey's uranium reserves were found to be approximately 10,000 tons. As these reserves will be sufficient to supply nuclear power stations for 30–50 years, Turkey will not be dependent on external resources of uranium when it establishes its stations. However, as the existing reserves have high costs associated with their extraction, the MTA has started to search for new reserves with costs below $80 per kilogram. In the latest surveys conducted by the MTA, new uranium reserves were found in Kırklareli, Balıkesir, Manisa, Kütahya, and Uşak. Following detailed studies, experts state that these reserves might be capable of providing uranium supplies for 100 years.

After the latest survey, an important uranium deposit with 0.05% tenor (the mineral content of an ore) was discovered in Köprübaşı, Salihli, and Manisa. In Köprübaşı, the MTA established a plant in 1974 to produce yellow cake (a mixture of different uranium oxides and other uranium compounds), and this plant was later transferred to Etibank and finally closed in 1982. A new survey, conducted in the same region with state-of-the-art technology, revealed one of the most valuable uranium deposits in Turkey. Previously, uranium deposits were discovered in the Sorgun district of Yozgat. The 0.1% U308 tenor uranium reserves are considered one of the richest uranium reserves in Turkey. The reserves here were estimated to be approximately 3850 tons.

New uranium deposits were detected at the bottom of Lake Van with a concentration of 0.1%–0.5% tenor. Furthermore, the coal ores in Elbistan district of Kahramanmaraş, the Beypazarı district of Ankara, the Orhaneli district of Bursa, and the Çan district of Çanakkale and the asphaltite ores in Şırnak were found to contain uranium deposits. The 0.04%–0.05%t tenor uranium reserves in Köprübaşı district of Manisa were estimated to be 2852 tons. The survey also discovered 490 tons of 0.05% tenor uranium reserves in Fakıllı located in Uşak, 208 tons of 0.04% tenor uranium reserves in Küçük çavdarlı located in Aydın, and 1720 tons of 0.08% tenor uranium reserves in Demirtepe in Ankara.

Historical review of uranium exploration

Uranium exploration in Turkey began in 1956 and was aimed at the discovery of deposits of vein-type uranium in crystalline terrain, such as acidic igneous and metamorphic rocks. Some pitchblende mineralization was discovered as a result of these activities, but they did not form economically viable deposits. Studies were undertaken in sedimentary rocks since 1960, covering crystalline rocks, and some small ore bodies containing autonite and torbernite mineralizations have formed across various parts of the country (Cameron, 1963). In the mid-1970s the first hidden uranium deposit with black ore below the water table was found in the Köprübaşı area and, as a result of more recent exploration, a uranium mineralization has been found in Neogene sediments in the Yozbat region of central Anatolia (Uranium, 2005). The recent and continuing discovery of uranium involved complex activities. In 2003–2004, a filed radiometric and geological prospection was performed in Central Anatolia. The results indicate no economically viable uranium ore is present in this area. Granite and acidic intrusive rocks and sedimentary rocks around the Sulkayurt (Ankara)-Kaman (Kirsehir) area will be searched for radioactive raw materials. For this purpose, a prospecting study throughout 3500 km^2 was carried out in 2006 (Uranium, 2005).

Uranium resources

The identified uranium resources in Turkey of the Reasonably Assured Resources (RAR) category reaching 9129 tU and occurring in the USD 80/kgU category (as in-site resources). These were reported from the following deposits (Uranium, 2005):

- *Salihli-Köprübaşı:* 2852 tU in 10 ore bodies and at grades of 0.04–0.05 U_3O_8, in fluvial Neogene sediments.
- *Faili:* 490 tU at 0.05% U_3O_8 in Neogene lacustrine sediments.
- *Kocarli:* 208 tU at 0.05% U_3O_8 in Neogene sediments.
- *Demirtepe:* 1729 tU at 0.08% U_3O_8 in fracture zone in gneiss.
- *Yozgat-Sorgun:* 3850 tU at 0.1% U_3O_8 in Eocene deltaic lagoon deposits.

The uranium occurrences in the Köprübaşı region (Anace et al., 1985, and Cameron, 1963) are situated 400 km WSW of Ankara on the main road from Demirci to Manisa on the northern side of the Demirköprü Dam. The uranium-bearing sediments are primarily gravels, siltstones, sands, and clays that cover the North Menderes Massive Palaeozoic gneiss and schists. There are also some calcareous breccia and limestone beds, as well as phosphates. The mineralized horizons in the sediments are not associated with any particular bed and differ considerably in thickness from 0.25 to over 10 m, with the average thickness being 4 m. The depth of the mineralized horizon from the surface also varies considerably, with an average thickness of 8 m, from outcrops of over 25 m.

The unconsolidated existence of both overburden and mineralized horizons will allow low-cost mining with no extensive explosive use. The secondary minerals of uranium (autonite, 122 meta-autonite and torbernite) bind or loosely cover individual grain surfaces. Pebbles or cracks are easy to leach and would require low-cost treatment by heap leaching. By 1963 a total of 65 surface trenches and 115 holes were constructed on a 50 × 50 m grid. Drilling totaled 4500 m and sunk five shafts for a total depth of 82 m. More research experiments under a program funded by the United Nations Development Program (UNDP) were carried out in 1974. Progress was achieved with the use of modern logging equipment for radiometric and electrical resistivity and conductivity measurements. To confirm earlier findings, more than 20 holes were drilled and logged in the deposit at Kasar. In 1978, MTA checked the estimates of the ore reserve.

Ore reserves

The results of earliest estimates of the ore reserve were published in 1963. MTA discovered additional ore reserves as further exploration was conducted at Kasar (the Kasar deposit is considered to have the largest reserve and a simpler amenability to the leaching process) and other ore bodies that culminate in the final geological reserves (Table 5.10.1). Table 5.10.2 contains the final figures for the reserve material.

The four deposits above were chosen for their suitability for open-pit mining. Reported incidents in the region were overlooked as they were sub-marginal for open-pit mining. The only exception is the Ecinitas deposit where, through underground mining, an appreciable amount of ore could be produced. It is assumed that the above reserves could supply

Table 5.10.1 Ore reserve estimation for uranium deposit in several locations in Turkey.

	Ore reserve			Waste	
Deposit name	Tons	% U	U tons	m^3	Stripping ratio m^3 ton
Kasar	**1,325,760**	0.0349	463	**7,954,560**	6
Cetintas	**811,000**	0.0258	209	**3,244,000**	4
Topalli	**577,800**	0.0329	190	**1,848,960**	3.2
Tasharman	**1,408,300**	0.0453	638	**12,956,360**	9.2
Total	**4,122,860**	0.0364	1500	**26,003,880**	6.3

Table 5.10.2 Mineable reserves of Köprübasi uranium deposit.

	Ore reserve			Waste	
Deposit name	Tons	% U	U tons	m^3	Stripping ratio m^3 ton
Kasar	**1,140,000**	0.036	416	**1,000,000**	0.9
Cetintas	**330,000**	0.027	91	**660,000**	2
Topalli	**310,000**	0.037	312	**470,00**	1.5
Tasharman	**510,000**	0.04	114	**1,900,000**	3.7
Total	**2,290,000**	0.04	933	**4,030,000**	1.75

mining and processing facilities with a mine ore run capacity of 230,000 tons producing about 80 tons uranium per year over the period of 10 years. Estimates of the MTA reserves were based on the polygonal approach using a minimum grade of 0.015% uranium without regard to its economics, feasibility, or physical mining warranty.

Ore types

Mineralization of the ore body according to mineralogy and lithology can be classified into three main types: (1) Kasar type, where uranium occurs in a fine matrix as autonite, meta autonite and torbernite; (2) Tasharman type that contains large quantities of phosphate as dahllite and apatite, with a greater amount of carbonate than Kasar and non-existent relevant mineral uranium; and (3) Carbonate type, which is distinguished by high calcite with some apatite dahllite.

Turkish Atomic Energy Authority (TAEK) and the MTA are still examining whether uranium reserves in Turkey would prove profitable or not. As the reserves have small amounts of uranium, the mining of these reserves may not be cost-effective. The reserves with mining costs lower than $80 per kilogram are deemed profitable, with mining costs of the reserves in Turkey estimated to vary between $80 and $130 per kilogram.

Turkey's thorium resources

Another raw material that can be used in nuclear power stations is thorium. One site of thorium ore in Turkey is in Kızalcaören in the Sivrihisar district of Eskisehir. Kızalcaören's thorium reserves are among the richest in the world, but there is currently no nuclear power station that operates on thorium. These reserves were discovered to be 0.2% ThO_2 and 380,000 tons. There are also thorium reserves in Kuluncak, Hekimhan, and Malatya. As the world's thorium reserves are concentrated within a limited number of countries, including Turkey, and there is currently no problem with respect to uranium production, no nuclear power station designed to operate using thorium has been commissioned. Experts argue that Turkey should revise its nuclear power station projects to be compatible for thorium. If Turkey establishes thorium-based nuclear power stations, it will never be dependent on external thorium resources.

CHAPTER 5.11

United Arab Emirates, Oman, Qatar, and Kuwait

Geological outlines

The Arabian Gulf countries include Oman, the United Arab Emirates, Qatar, Bahrain, and Kuwait. These countries form the eastern and southeastern marginal zones of the large Arabian Basin. The Basement Precambrian rocks form a small percent of the exposed rock, while the rest of the exposures belong to the Phanerozoic sedimentary sequence. In general the older rocks exposed in the south and southeastern parts of the Arabian Gulf, while the rocks becoming younger to the north.

The most significant outcrop of the oldest rocks in Oman is in Dhofar, east of Marbat and on the Hallaniyat islands. They are granite clan crystalline basement blocks, gneisses, and mica schist. Granodiorite also outcrops on the southern edge of the Wahiba Sands in a small area near Jabal Ja'alan. Late Proterozoic to Cambrian rocks outcrop in the al-Huqf on the eastern side of the al-Wusta area just south of the turn toward Masirah Island, and on Saih Hatat south of Muscat (Fig. 5.11.1). These have a predominantly terrigeneous and volcanic origin, characteristic of continental shelf environments, with subsequent transgression to a marine climate.

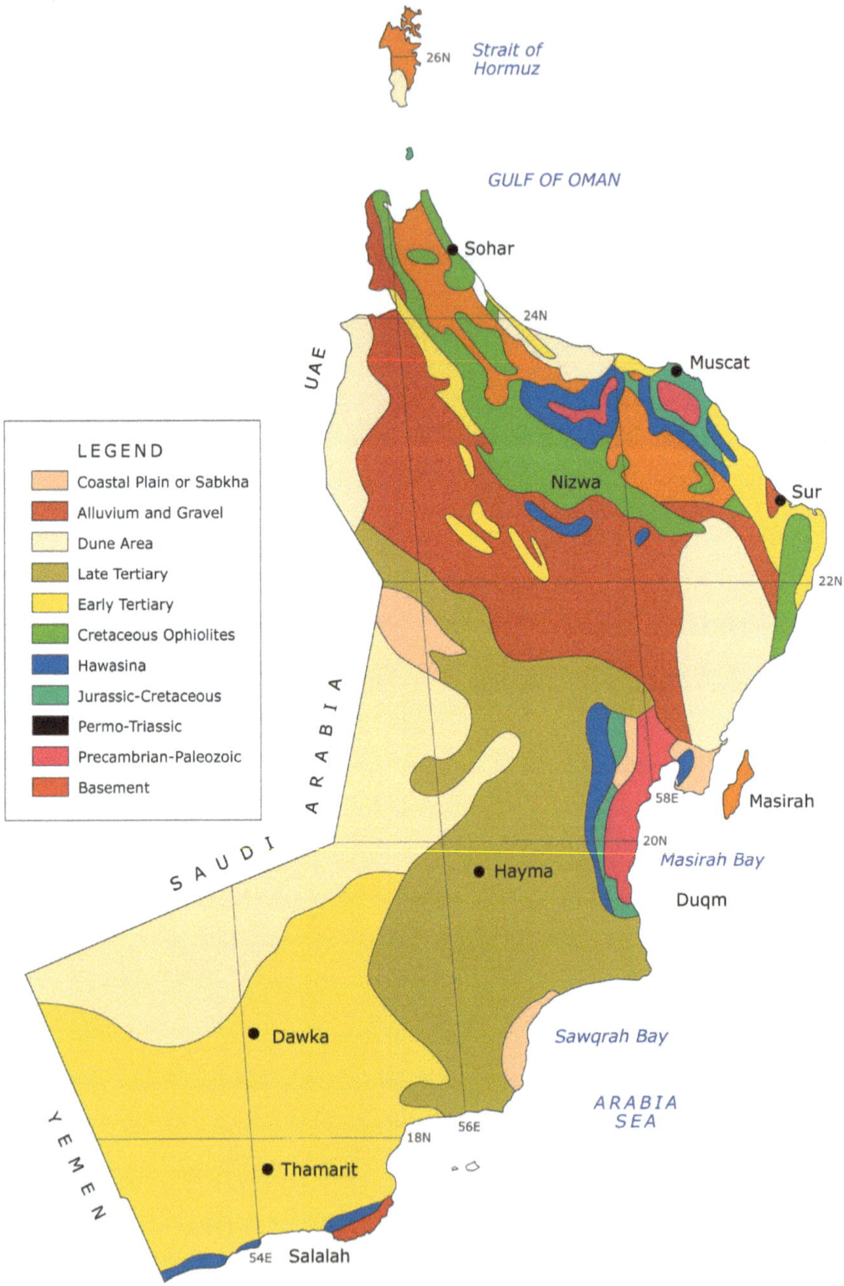

Fig. 5.11.1 ***Simplified geological map of Oman.***

It is clear from the simplified geological map of Oman (Fig. 5.11.1) that the old rock sequences and the younger rock exposures are often controlled by tectonics of the region. For example, the older rocks outcropped in the northeast of Oman are separated by Early Tertiary to Recent sediments from the al-Huqf Precambrian-Palaeozoic exposures. There should be a tectonic basin following NW-SE direction, which mostly follows the trend in the Arabian Gulf. Also, the outcropping of Basement rocks and old Palaeozoic sediments follow the E-NE trend. These exposures are often controlled with ENE-WSW trending faults, parallel to the Arabian Sea. Following the rock exposures from Oman to the United Arab Emirates, no basement rock exposures can be seen. The exposed sedimentary sequence only includes Tertiary and older sedimentary rocks in addition the quaternary to recent sediments (Fig. 5.11.2).

Geologically, the United Arab Emirates can be divided into two distinct parts: (1) the desert floor and coastal plains dominated by evaporite flats and sabkhas, as well as gravel plain and aeolian sands, and (2) the mobile Oman Mountains, situated at the eastern end of the Arab plateau and flowing abroad parallel to the Gulf of Oman. The Arabian Gulf also includes offshore islands, such as Dalma and Sir Bani Yas. Such islands are made up of rocks much older than those found in the Oman Mountains. The islands are almost circular in shape and this is due to their origin as Cambrian salt diapers. Also, there is a salt dome structure on the mainland at Jabal Az Zannah. The following sites represent some of the important geological exposures in United Arab Emirates.

Jabal Hafit

Jabal Hafit is situated in the city of Al-Ain to the southeast of Abu Dhabi. It is about 29 km in length and 5 km wide, with an elevation reaching 1160 m above sea level. It is bounded by the city of Al-Ain to the north, by the plain of al-Jaww to the east, by Mazyad and Oman to the south, and by Ain al-Faydah and Zakher to the west. It is an eroded anticline (Figs. 5.11.3 and 5.11.4), with beds dipping eastwards and westward on the limbs. The overall morphology gives it the feel of a "whaleback." It is also believed that the source of the famous hot springs is believed to be located beneath Jabal Hafit. Jabal Hafit has the most complete sequence at the United Arab Emirates to review Eocene and Oligocene rocks.

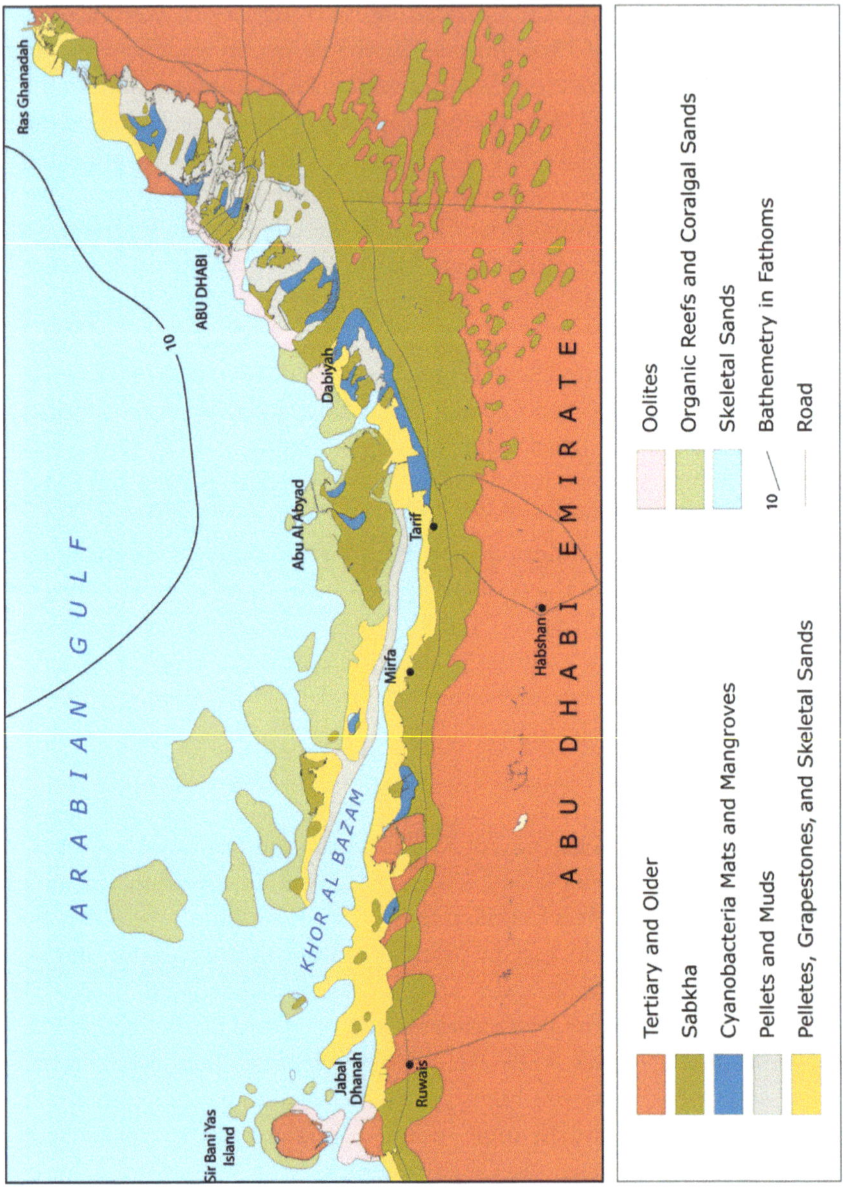

Fig. 5.11.2 ***Simplified geological map of the United Arab Emirates.***

Fig. 5.11.3 ***Western limb of Jebel Hafit anticline, showing the water bond.***

Jabal al-Rawdah

Jabal Al-Rawdah is situated at the northwestern end of the Hatta Zone, a 50-km-long WNW narrow region of compressional and sinistral shear deformations. The two most prominent topographical features within and bordering this region are the Jebel al-Rawdah-Masfut Ridge (14 km long and parallel to the Hatta Zone) and Jabal Quimah, situated in the south-eastern part of the zone. Most of the structures in the Hatta Zone and at Jabal al-Rawdah may be linked to the Semail Ophiolite obduction over the continental margin. This occurred about 90 million years ago, at the end of the Cretaceous Period.

Fig. 5.11.4 ***Part of the Jebel Hafit anticline, showing the hot water spring flushing.***

During the Oligocene, the Jabal al-Rawdah sheared rocks and their carbonate cover were later refolded. These folds are large "whaleback" structures (like Jabal Hafit) that trend from NNW to NNE. The nearby Jebel al-Faiyah mountain is a large fold of this group's NNE trending. However, the fold pattern at Jabal al-Rawdah is WNW, parallel to the Hatta line. This is interpreted as being because of zone reactivation.

Jebel al-Faiyah

This formation is located at about 90–100 km from the town of al-Ain, on the west side of the Oman Mountains. The al-Faiyah mountain range is an eroded anticline with exposed core of weathered and oxidized ophiolitic rocks. The neoautochthonous sedimentary rocks in the Al-Faiyah Range are the Mastrichtian Qahlah and Simsima, and the Eocene Dammam formation. Such units unconformably overlie the allochthonous Semail Ophiolite.

State of Qatar

Further to north, the state of Qatar comprises 11,437 square kilometers of flat limestone as a peninsula stretching from a western base roughly 180 km north into the Persian Gulf. This base separates the Saudi Arabia and Abu Dhabi coastlines. The peninsula lies along the crest of the Qatar Arch, a tectonic structure that stretches almost north–south to the Gulf and divides it effectively into two basins. The Qatar Arch has an important impact on Gulf's tectonic patterns and geology since the Paleozoic Era.

Kingdom of Bahrain

Bahrain is a generally flat, consisting of a low desert plain that gently rises to a low central escarpment. The highest point is the Jabal al-Dukhan, 122 m high. Bahrain's total surface area is 688 square kilometers and comprises an archipelago of 33 islands located midway through the Arabian Gulf near the Arabian Peninsula's shores. The total area of these islands is approximately 691 square kilometers. Bahrain, with an area of 572 square kilometers (Fig. 5.11.5), is the largest, accounting for 83% of the total area.

Surrounding much of Bahrain is a fairly shallow Arabian Gulf inlet known as the Bahrain Gulf. The island is largely low-lying and barren desert. Low rolling hills, stubby cliffs, and shallow ravines create outcroppings of limestone. The limestone is covered by varying saline sand densities, capable of supporting only the harshest desert vegetation—mainly thorn trees and scrub. The interior includes an escarpment that rises to 134 m,

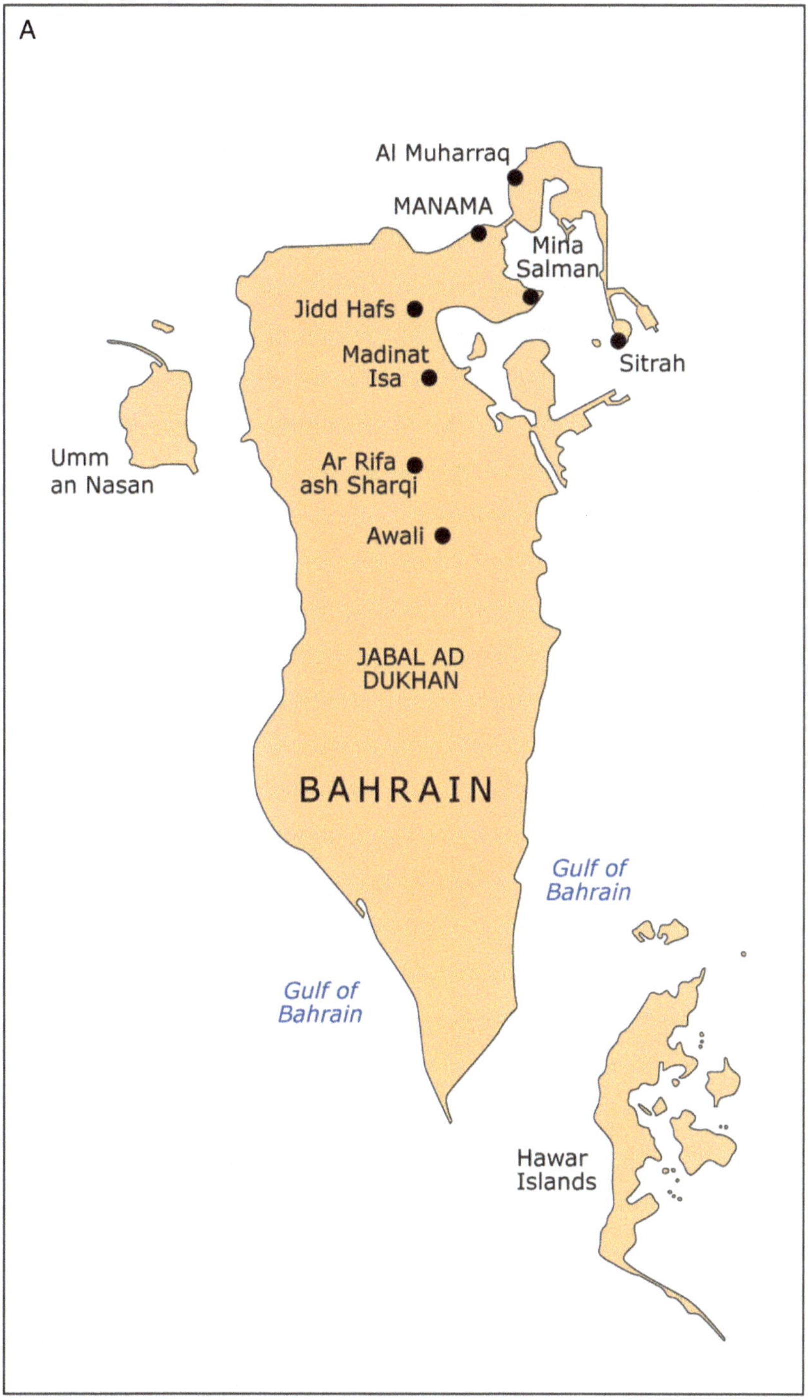

Fig. 5.11.5 ***Map and satellite image of Bahrain.***

Fig. 5.11.5 *(Cont.)*

the island's highest point, to form Jabal al-Dukhan (Mountain of Smoke), named for the mists that sometimes wrap around the summit. Most oil wells in the country are in the vicinity of Jabal al-Dukhan.

Kuwait

Kuwait lies in the far northwest corner of the Arabian Gulf. It is a small state of approximately 17,818 square kilometers and only about 200 km north to south at its furthest points, and 170 km east to west. It is bounded by Iraq in the north and west, Saudi Arabia in the south, and the Arabian Gulf in the east. The State of Kuwait includes the mainland and nine offshore islands, including the inhabited Failaka island, the low-lying muddy Bubiyan island near the Shatt al-Arab mouth, and seven small coral islands mainly in the south.

The land was formed in a recent geological period. Limestone rises in a long, north-oriented dome below the surface in the south. The main oil fields, Kuwait's most important natural resource, are located within and beneath this formation. Layers of sand, gravel, silt, and clay in the west and north overlap the limestone to a depth of more than 210 m. The upper portions of these beds are part of a mass of sediment deposited by a broad wadi whose most recent channel was the Wadi al-Batin, the deep shallow valley that forms the country's western boundary.

Potential uranium resources

No real uranium resources have been reported in any literature of the Arabian Gulf countries. However, according to the geological criteria of some of these countries, some potential uranium sources are possible. In Oman, the geological and structural features are very suitable for hosting uranium deposits. The pan-African younger granite can form source and host rock for uranium. The zones of intersection of ENE-WSW and NW-SE faults and shear zones form good traps for uranium mineralization. Moreover, the contact zones with other country rocks often form promising zones for the localization of uranium. The basal conglomerates at the contact zones between the Precambrian rocks and the Palaeozoic sequence can form promising sites for hosting uranium mineralization. Moreover, the phanerozoic sedimentary column should be investigated, especially the Carboniferous sequence. The phosphate-bearing sediments and the associated structures should be considered as targets for uranium exploration. It is important to re-interpret the radioactivity records of oil and water wells, especially in United Arab Emirates, Qatar, Bahrain, and Kuwait. This will help to define some uranium-bearing horizons. This is of prime importance due to the critical nature of uranium resources for any peaceful nuclear programs for generating power and the desalination of sea water. Also, it is essential to consider that oil is a limited fossil fuel that will cease to exist in the coming decades.

CHAPTER 5.12

Yemen

Geological outlines

The Republic of Yemen is situated on the southern tip of the Arabian Peninsula covering an area of 527,870 km^2 and with a population of about 21 million in 2005 (Fig. 5.12.1). Geologically, Yemen forms a part of the Arabian Shield. This shield forms the East Africa–Arab craton's eastern portion. The rocks at the base are some of the oldest structures dating back about 3 billion years. These structures reflect magmatic rocks, gneiss, and schist rocks. These structures appear as belts that are hundreds of kilometers long as ancient rock exposures separating the small sheets that joined to form the Arab–African Shield.

Four gneiss lands and two island-arc terrains are found within Yemen's Precambrian area. In Saudi Arabia, the two western gneiss terrains can be correlated with well-documented lands (Asir and Afif). The terrain in Yemen correlates with comparable terrains on the Arabian Shield's eastern margin and in northern Somalia. Thus, Yemen provides a link between the arc of the Arabian Shield and East Africa's gneissic Mozambique belt. Yemen's oldest rocks of sedimentary structures date back to the prime Proterozoic period. These rocks are characterized by groups of Ghaba'er and Gunab with sandstones and limestones. Sedimentary rock types–the sandstones, limestones, and the mudstones–cover large areas that form Yemen's surfaces and basins. Most of these rocks are geological of the second and third ages. Exceptions are the Wajid sandstones and the A'kberah mudstones, both of which are found in Yemen's western-northern region (i.e., Sa'adah) and were formed in the first geological period.

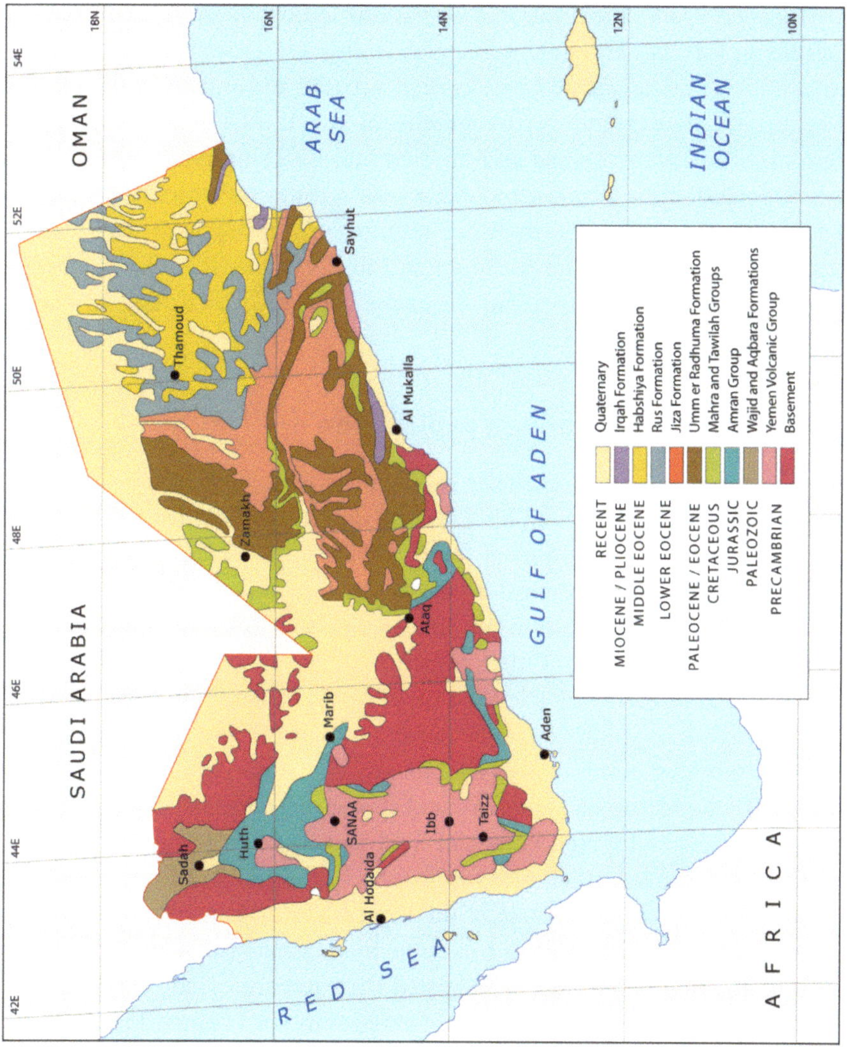

Fig. 5.12.1 ***Geological map of Yemen (modified after Petroleum Exploration and Production Authority (PEPA), Republic of Yemen (2008)).***

Yemen has significant exposure to volcanoes. The formation of the Red Sea system in the third Era was correlated with the early Eocene Epoch of the regional uplifting of Yemen's western areas. This was achieved through volcanic interval; the highest of which was in the epochs of the Oligocene and the Miocene. At the beginning of the fourth period, the volcanic eruption happened again. Such volcanic operations led to unique Yemeni volcanoes being formed with acid rocks such as rhyolite, Ignimbrite; volcanic glass, as well as middle and basic rocks such as basalt and andalusite. During the volcanic activities and in the middle of the third eon, overlapping granite structures were formed and injected into different rock types, most of which included the A'mran and Altaweelah groups of Yemen's volcanic and basement rocks. Yemen's structures are often controlled with two main trends, the ENE-WSW (Gulf of Aden) and the NW-SE (Red Sea). These two main structural segments can be easily delineated according to the nature of the exposed rocks.

Historical review of uranium exploration

Yemen is one of the Arab countries that has the least amount of available data about uranium resources. This is mostly due to the fact that only limited work has been carried out on the exploration of nuclear materials in general and especially uranium. This is due to several factors, notably the lack of facilities for exploration, in addition to the lack of foundational attention given to this area. Other priorities in development have dominated relative to the area of nuclear science. Nevertheless, there are some indications of the existence of radioactive anomalies that may reflect the possible presence of uranium ores in Yemen.

In 1987 the company Total, in collaboration with the Geological Survey of Yemen, conducted visits and field surveys to search for radioactive elements in the basement rocks, especially in the areas north of Saadah, southeast of Taez, and to the east of the capital Sana'a. It is suggested that it is the most encouraging areas to search for nuclear raw materials. The results obtained by the company indicated the presence of concentrations of radioactive elements in the crust of ferruginous sand stones. This work also discovered radioactive anomalies in pegmatite veins especially those in the southeast of Taizz. It is worth mentioning that the studies reported there is no economic feasibility of these elements at the present time and recommended further research and exploration to detect other sites with economic value.

Areas of radioactive anomalies

These following areas are in the province of basement rocks and will be treated according to the results of research conducted to date.

Al-Gebn (southeastern region of Marib)

This village is located 50 km south of Radea city. This site was discovered by chance by the citizens and inhabitants of the area who brought rock samples that gave up to 1000 ppm uranium after analysis. The chemical analyses done by the laboratory of the International Atomic Energy Agency indicated the presence of uranium, thorium, and actinium. The Geological Survey of Yemen sent a team to visit the site and to perform some geological and radiometric foot prospecting. The radiometric measurements varied from 500 to 16000 cps particularly in the locations of granite exposures that are rich in magnetite.

Alhamorh

The Alhamorh area is located about 40 km to the southeast of Taizz. Geologically, it is covered with highly metamorphosed pre-Cambrian rocks. The Geological Survey of Yemen started prospecting for copper and nickel mineralization, and during exploration some studies were conducted to explore and assess the nature of the radioactive anomalies in the pegmatite in the area. The results of the studies showed the presence of element concentrations (uranium and thorium) in varying degrees, in addition to the existence of many rare elements (lanthanum and yttrium). The studies also showed the presence of radioactive anomalies in the vicinity and some valleys such as Wadi Arar, Alaabbos, and Alshoweifa.

Sa'dah

Sa'dah city is located 200 km north of the capital region and covered with pre-Cambrian rock exposures. The area gained considerable attention from the state and researchers due to the existence of mineral deposits of economic value such as iron, copper, tin, gold, and others. The Geological Survey of Yemen, in cooperation with foreign companies, is implementing several exploration activities and geological projects in this area. During these activities, some discovered phenomena include encouraging uranium deposits. The preliminary studies indicate the existence of radioactive

Table 5.12.1 Results of the analysis of rock samples.

Sample number	Cu	Bi	Fe	Ni	Pb	U	Zn	Co	Cr
No.	ppm	ppm	%	ppm	ppm	ppm	ppm	ppm	ppm
R–1	2	30	3.36	18	217	148	792	35	6
R–3	3	21	1.74	10	130	81	39	5	39
R–5	3	45	7.5	47	145	127	2100	63	16
R–6	2	30	6.6	40	66	122	2160	26	36
R–7	1	15	4.74	15	206	1031	41	20	5

anomalies and radioactivity indicators within the following geological units:

1. Granite rocks and pegmatite veins of the pre-Cambrian.
2. Sandstone rocks of Ordovician to Permian age which belong to the sedimentary cover, and show exposures in limited areas north of Sa'dah.

These rocks are coarse sand grains partly ferruginated and are often like the Nubian sandstone deposits in Egypt and Sudan, which are characterized by nonconformity with the basement rocks. This geological environment is the most likely area to host sedimentary uranium deposits. The results of chemical analyses for uranium and other elements (Table 5.12.1) in rock samples showed relatively high ratios of uranium content in the rocks.

Maoiwiah

This area lies about 25 km east of Taizz. The main exposures are formed of Yemeni volcanic rocks. This site has been known from the radioactivity maps of the joint project of Natural Resources which was performed in 1984–1985. The results show a high concentration of radiation measurements at the locations of some acidic volcanic dykes that intersect older rocks such as granite plutons.

Al-Sobeiha

This area is located southeast of Taizz and is geologically covered with pre-Cambrian rocks. Some of the information and reference sources of the Geological Survey of Yemen indicate the improved presence of radiometric anomalies in this region.

Al-Jawf (northeast)

This area lies northeast of Sana'a and is covered with basement rocks that represent an extension of the basement rocks of Sa'dah. This area is considered as one of the promising areas for important metallic deposits. During search operations and mineral exploration carried out by various small working teams, some radiation readings were recorded with varying values. These relatively high radioactivity readings are hosted in the granites and the pegmatite rocks in the area; however more work is needed to evaluate their importance.

Potential uranium resources in Yemen

Based on the review of the available data on the geological phenomena and the recorded radioactive anomalies and their host rocks in various areas, Yemen could hold rich uranium resources. It is essential that the compilation and classification of all the data obtained during the previous explorations for uranium in Yemen commence as soon as possible, putting these data on a base map to determine the geological formations that could contain uranium deposits. It is important to study and reevaluate the water and oil well logs to studying the geological column and radioactivity changes within the geological column of Yemen. This could lead to important uranium resource discoveries in Yemen.

The old basement rocks form remarkable exposures in Yemen and conglomerate-type uranium deposits could be defined in these Proterozoic sequences. Also, the metamorphosed Proterozoic sandstones are often good sources for potential uranium. The Pan-African granites represent good targets for hosting uranium resources, particularly the two mica types. The contact zones of these granites can be considered as good potentialities for uranium resources. Moreover, the intra-cratonic basins within these granites, which include metamorphosed rocks, are considered as good targets for uranium exploration, as well as the basal conglomerates that separate the basement rocks from the younger sedimentary sequences, especially the Palaeozoic sedimentary rocks. This phenomenon was recorded in Egypt (Salman, 2005). It is important to note that the Yemen volcanoes (Figs. 5.12.2 and 5.12.3) especially the acid types can be considered a good source and host for economic uranium deposits (Volcanogenic Uranium Type). The associated geological structures within or nearby these volcanic areas can be considered as good targets for uranium exploration.

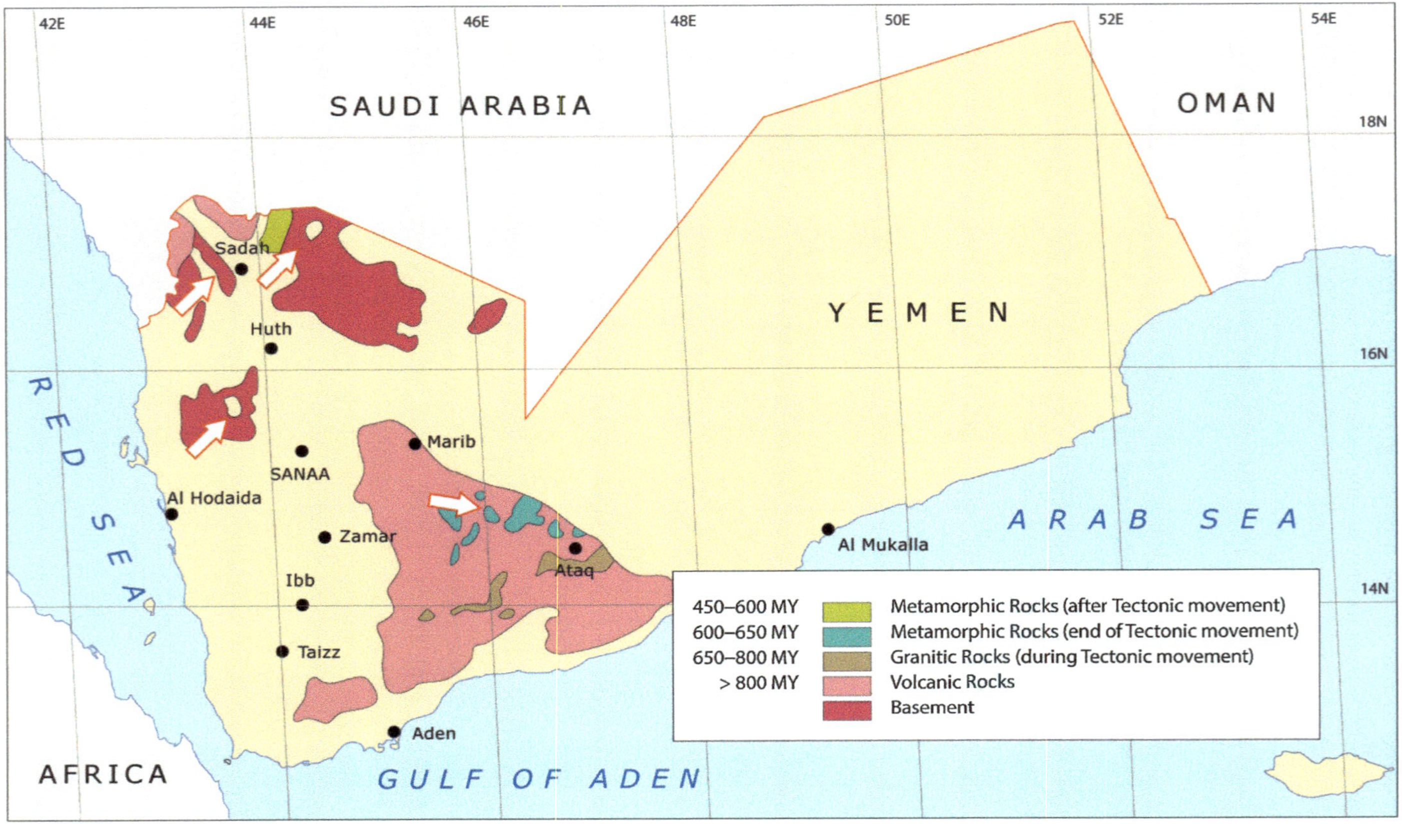

Fig. 5.12.2 ***Map shows the types of basement complex, showing granite rocks as potential uranium provinces (indicated by arrows).***

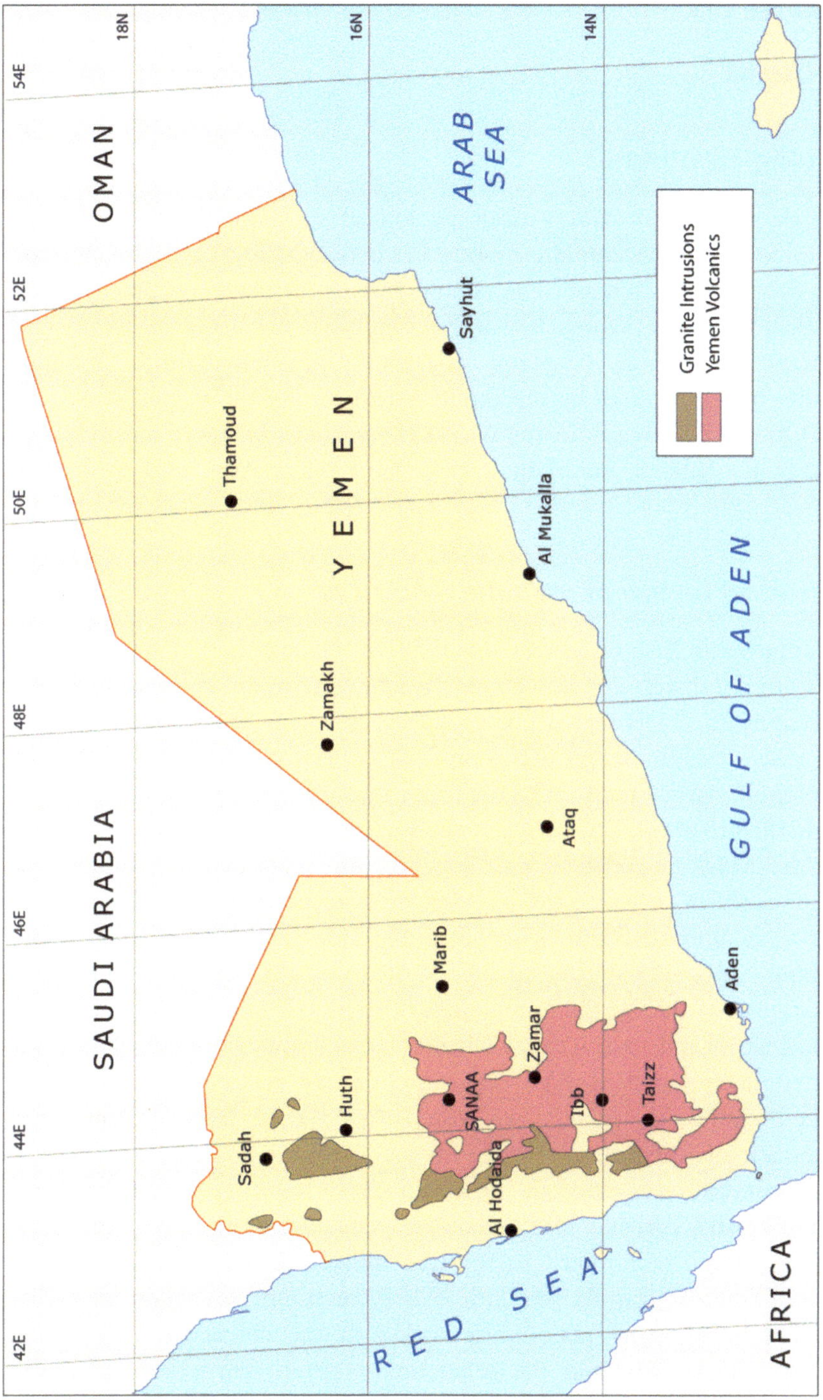

Fig. 5.12.3 *Map shows interfingering of the granite rocks and Yemen volcanics as potential uranium provinces.*

CHAPTER 5.13

Algeria

Geological outlines

Algeria is covered with various geological units ranging in age from Precambrian to Recent. The country lies between the West African Craton and Saharan Meta-craton. The central part of Algeria is occupied by the Saharian Platform, which is mainly composed of Paleozoic sedimentary sequences with some Cenozoic volcanism. This central part is bounded mainly from the south by the Targui Shield and by the Regui Shield and the Ougarata Range from the West. The northern parts are mainly occupied by the Saharan Atlas Highlands and the Tellian Atlas (Figs. 5.13.1 and 5.13.2).

Historical review of uranium resources

Uranium was first discovered in 1958 in Huajjar near Tmguin, about 200 km southwest of the city of Tamanrasset by an exploration group belonging to the French Geological Survey. The features associated with the discovered uranium occurrence have been studied by the French Geological Survey and later by the French Atomic Energy Authority (CEA) from 1958 to 1960. The CEA has conducted some subsurface explorations studies, which included gamma logging for the drilled bore holes to a depth of 50 m, where many exploratory bore holes were made. In 1968, two years after founding of the National Company for Exploration and Mining (SONAREM), experts studied the criteria associated with uranium in Tmguin. Moreover, the bore holes drilled by CEA at greater depths have been logged and radioactive anomalies were found. From 1968 to 1970, the company (SONAREM) began preliminary exploration, where they started to study the mineralized areas. Moreover, the company conducted systematic radiometric surveys for the basement exposures.

Those efforts resulted in the discovery of uranium deposit basins in Tmguin, Abankor, and Tiniv (about 50 km from Tmguin). At the same time, a ground radiometric survey was initiated in the sedimentary rocks surrounding Hoggar. At the end of 1970, uranium anomalies were discovered by airborne radiometric survey at the contact zone between the basement rocks and the sedimentary cover, which belong to the

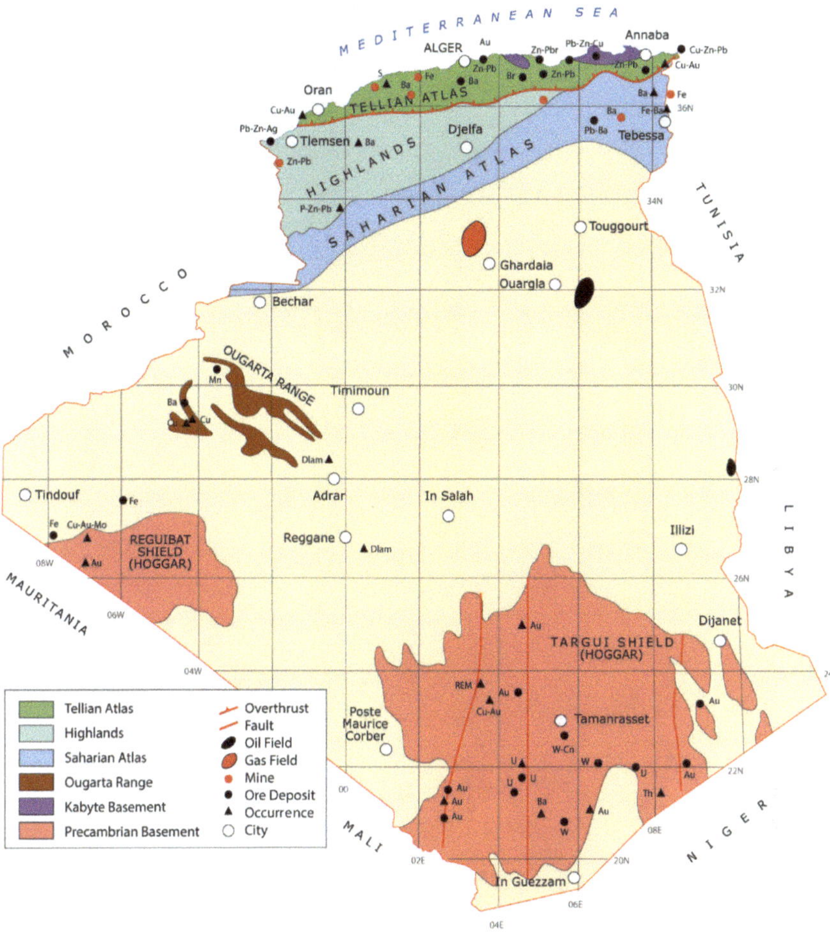

Fig. 5.13.1 ***General geological map of Algeria showing some uranium prospect sites with modifications.***

Cambro-Ordovician ages and in the lower horizons of the Devonian sediments. The processes for uranium exploration and evaluation belonged to SONAREM, which has allocated a part of its development program to the Tamanrasset State. The national program for uranium exploration included the following activities:

- Systematic exploration in Central, Eastern, and Western Huajjar.
- Ground proving for radioactive anomalies discovered by airborne radiometric survey.
- Prospecting to increase uranium resources in the Tmguin, Abankor, and the Tin–Sudairin Basins.

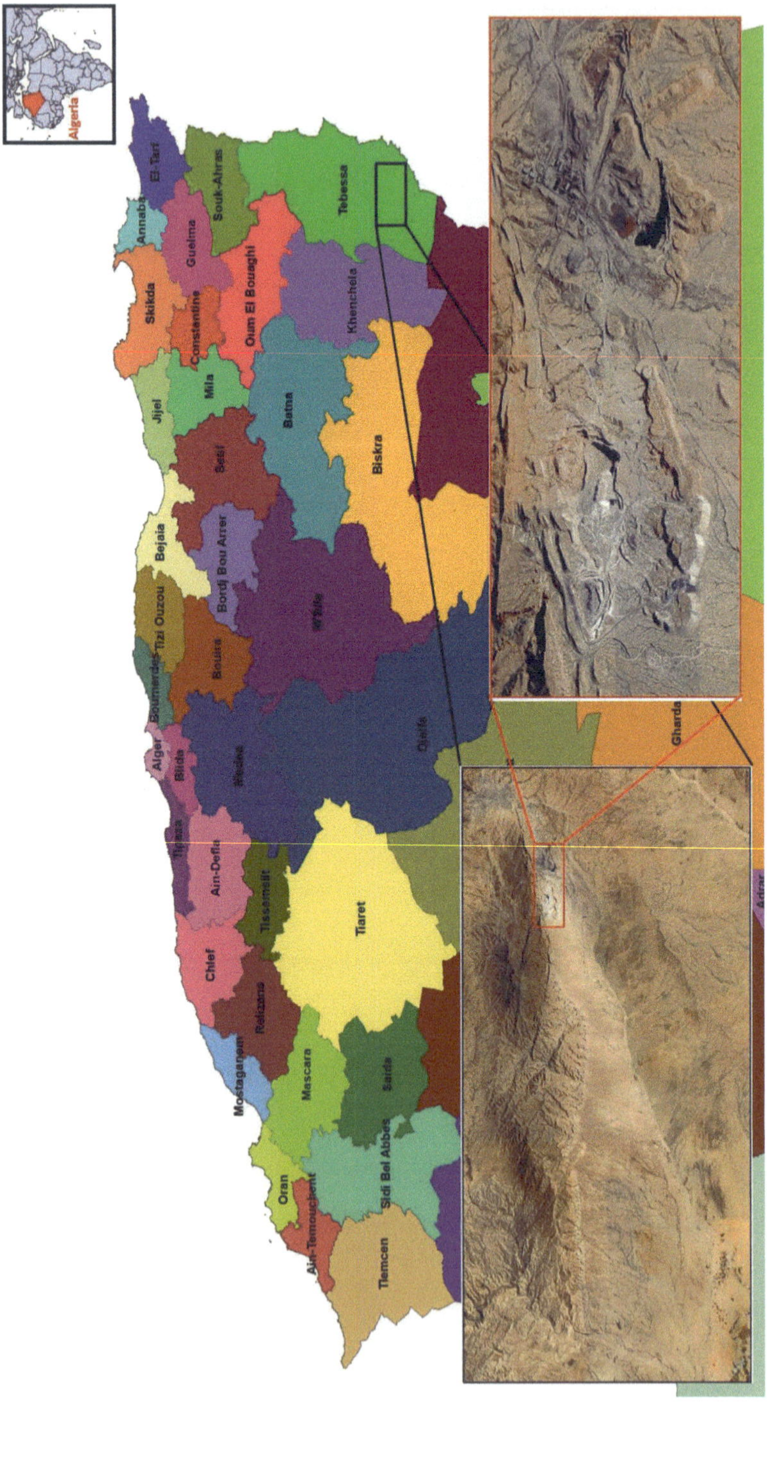

Fig. 5.13.2 *Phosphate open mining in Tebessa Basin, Algeria.*

Uranium prospects

The main uranium resources (Table 5.13.1) with estimated reserves are present in Alehaugar, southern Algeria in Tmguin, Abankar, and Teneif, about 200 km south west ofTamanrasset.The Palaeozoic rocks in Alehaugar are divided into the following two groups:

- *Zogadian group:* It mainly consists of metamorphic rocks (gneiss, quartzite, schist, and marble).
- *Alvarosean group:* It consists mostly of metamorphic volcanic rocks.

It should be noted that all these rocks are affected by regional faults and shear zones following a northsouth axis. These faults delineate the Suggarian heights (Suggarian Horsts) and Farozian Grabens.The uranium deposits are hosted in the Recent Plutons near the contact points of these structures. The Tmguin pluton is mainly composed of alkali double mica granite.This granite pluton has been affected by a series of faults that strike in the direction of the North 5-15 East, which dislocate its eastern edge. Secondary fractures strike at the North 40-50 East directions and intersect the previous fault set and the same eastern edge. The uranium deposits in the three sites; central Tmguin, Abankar, and Tenief are in and around these primary faults (Salman, 2005).

The uranium mineralization belongs to the vein in these three sites, where they vary in grade. In Central Tmguin, the uranium content is 1% within the veins and up to 0.1% in the stock work. In Abankar, the average content of uranium in the ore is higher than that in Tmguin, Tenief, or Abankor. According to available data of the Red Book (El Aassy, 1984) in 1982, exploration work has resulted in the discovery of the following deposits:

- Four vein-type uranium deposits in Tmguin, Abankor, Dera, and Tenief. These deposits are hosted in migmatitic gneiss that belongs to Alsojoran and Alfarosean. The geological situation illustrates that they a have a Precambrian horst, which is cut by units of Quartzite, Pyroxenite, and amphibolites. Sometimes this horst is partly covered with volcanic material composed of rhyolite and breccias. The primary uranium minerals are present in the form of pitchblende and uraninite, with secondary uranium minerals such as Autonite, torbernite, uranophane, etc. While the secondary minerals are present as sulfides of lead, zinc, copper, iron, molybdenum, in addition to hematite and magnetite.
- Uranium-bearing beds in Tasees South Alehaugar in the Tin–Sreden Basin. At this site the basement rocks are covered with Sogarian Conglomerate and limestone, which belong to Cambrian-Ordovician.

Table 5.13.1 The main characteristics of uranium resources in Algeria.

Uranium deposit type	Characteristics
Vein-type uranium deposits	**Hosted in Pan-African granites:** The vein-type uranium deposits are found in veins in primary fractures linked to faults around granite batholiths. This form of deposits comprises the deposits of Timgaouine, Abankor, El-Bema, and Ait Oklan deposits in the south of Hoggar. The reserves of vein-type uranium (RAR) are 24,000 tons (Uranium, 2005). The primary uranium minerals are present in the form of pitchblende and uraninite with secondary uranium minerals as autonite, torbernite, uranophane, etc. The secondary minerals are present as sulfides of lead, zinc, and copper, iron, molybdenum, in addition to hematite and magnetite (Type localities: Timgaouine, Abankor, El-Bema and Ait Oklan, Dera and Tenief, S. Hoggar).
Nonconformity-related type	**Basal conglomerates between Upper Proterozoic and Paleozoic:** Uranium-bearing beds in Tasees South Hoggar in Tin-Sreden Basin. At this site, the basement rocks are covered with Sogarian Conglomerate and limestone, which belong to the Cambrian-Ordovician Periods. These sedimentary rocks are conformably covered with Silurian shale and Devonian conglomeratic beds. There are concentrations of uranium hosted at the contact zone between the basement rocks and the sedimentary cover in the form of semi-horizontal bodies in the old depressions within the basement rocks. The thickness of uranium deposits ranges between 1 and 8 m. The small deposits in Tahagart have been evaluated. Uranium mineralization consists mainly of autonite and torbernite. The stratified-type uranium deposits (RAR) reach 2000 tons (Uranium, 2005), (Type localities: Tasees S. Hoggar in Tin-Sreden Basin).
Unconventional	**Phosphorite deposits:** Extensive phosphorite deposits occur in Algeria. The actual reserves are assessed at 2 billion tons, located in the phosphate basin of Djebel Onk (Tebessa). These reserves make up only a fraction of the phosphate basin (Int.Alg.1) Prévot (1990) mentioned that Algerian phosphate resources are 500 million tons with an average P_2O_5 of 25%. The average uranium content in these phosphorites is 0.009% U_3O_8. The uranium could be extracted as a byproduct during the manufacture of phosphoric acid (de Voto and Stevens, 1979), (Type localities: Djebel Onk (Tebessa).

These sedimentary rocks are conformably covered with Silurian shale and Devonian conglomeratic beds. There are concentrations of uranium which are hosted on the contact zone between the basement rocks and the sedimentary cover in the form of semi-horizontal bodies in the old depressions.

Based on exploratory studies conducted until 1983, the proved (RAR) uranium ore reserve in Algeria is estimated to be about 26,000 tons at costs less than 80 dollars per kilogram. The vein-type uranium deposits are 24,000 tons, while the stratified-type uranium deposits reach 2000 tons. As for the uranium resources in Algeria, reserve estimation of the Estimated Assured Resources category consists of two geological types: Upper Proterozoic non-conformity deposits and vein deposits (Uranium, 2005). The first type includes deposits linked to weathering profiles (regoliths) and deposits associated with the sedimentary cover basal conglomerate and sandstone. It is situated primarily in the south Hoggar basin of Tin-Seririne. Deposits of the vein type are associated with primary fractures within granite plutons in the fault zones. This form includes the South West Hoggar deposits of Timgaouine, Abankor, El Bema, and Ait-Oklan. Algeria reported a presence of RAR of 26,000 tons uranium costing less than $US 80 and 26,000 tons uranium costing less than $US 120 (Uranium, 2005).

Potential uranium resources

Reviewing the available information on uranium deposits in Algeria and the associated geological criteria, several geological environments are good potential sites for hosting uranium resources.

Pan-African younger granites and associated volcanics

The younger granite exposures can be considered as good target for uranium exploration in Algeria where it is proved to host some economically viable uranium deposits. The double mica granite in particular has a strong potential. The geological structures in the Basement Precambrian rocks, particularly the younger Pan-African granite and double mica granite and the acidic volcanic rocks, tuffs, and breccias are very promising targets.

Intra-cratonic basins

These structures show great potential for trapping uranium mineralization. Their pale-geographic characteristics, the nature of the surrounding basement rocks, and the associated geochemical phenomena are controlling factors for the localization of uranium mineralization. The geological structures in the Basement Precambrian rocks, particularly the younger Pan-African granite and double mica granite and the acidic volcanic rocks, tuffs and breccias. The intra-cratonic basins represent one of the important targets that should be explored carefully, as these structures could host an economically viable deposit of uranium ore in the vicinity of the radioactive anomalies. Attention should be given to the interface between the firearms of the rock era before life and the sedimentary cover and special scale, which contains Alesis Conglomerate. In this regard, basins in the synthetic sequence sedimentary rocks near those firearms represent an important target for the exploration of uranium. They represent important reservoir basins of uranium, especially if they are rich in organic matter and circumstances permit the geochemical concentration of uranium. The rocks carrying firearms are a main source trapping uranium, which could feed those basins containing a synthetic sedimentary sequence.

Lower Palaeozoic conglomerates

The contact zone of the Basement rocks and the Lower Palaeozoic sequence, especially basal conglomerates, is a potential target for uranium exploration. In this regard, the tectonic basins that include the Palaeozoic sedimentary rocks have good potentialities for hosting uranium deposits, especially if they are rich in organic matters and their geochemical characteristics are suitable for uranium deposition and fixation (Salman, 2005). The nearby basement and volcanic rocks represent the source that can feed these basins and form good traps for uranium. Moreover, the major shear zones and faults that dislocate both basement and sedimentary sequence in these basins can form passages or channels for moving uranium-rich solutions into these basins.

Phosphate deposits

Actual reserves are evaluated at 2 billion tons, localized in the Djebel Onk phosphate basin (Tebessa), and can be developed by open cast way with a maximum stripping ratio of 1/4. These reserves represent only a part of the phosphate basin.

CHAPTER 5.14

Libya

Geological outlines

Libya, situated on the Mediterranean foreland of the African Shield, extends over a platform of cratonic basins. The Precambrian basements rocks are mainly exposed in the Tibesti Mountains in the southern part of the country. The rest of the country is covered with sedimentary sequence ranges that vary in age from the Paleozoic to the Quaternary Periods. Tertiary volcanic rocks are widespread in al-Haruj, al-Aswad, and Gebel Nuqay (Goudarzi, 1978).

Rocks of the Precambrian basement are outcrop in many parts of southern Libya. The largest of these define three litho-tectonic provinces of distinctive rock assemblages and modes of deformation. The provinces are: the al-'Oweinat Basement Province (OBP) in the southeast, the Tibesti Basement Province (TBP) in the south, and the Ghat Basement Province (GBP) in the southwest. However the former is the oldest and encompasses Archean rocks, the latter two are much younger and are of mostly Pan-African age. All basement provinces encompass both metamorphic and igneous rock suites. Although metamorphism is prevalently regional, there is evidence of limited contact and retrograde metamorphism in certain locales. The metamorphic terrain is pierced by igneous bodies of different forms and sizes displaying granitic to gabbroic mineralogy. Although not a basement component, Tertiary ring intrusions of oversaturated to undersaturated composition add to the diversity of the igneous rocks of the al-'Oweinat region.

The OBP is of the Upper Archean to the Upper Proterozoic Age (2700–750 MA) with the latter being confined to diabase dykes transecting a largely peneplaned terrain. This province is composed of three structural domains, separated apparently by deep seated faults now buried under sand dune belts. Tertiary ring complexes piercing through the basement add to its tectonic framework, although subtly. The structural style varies from tight isoclinal flow-folding in the northern domain to open, flow-flexural folding in the southern domain. Only in a few places, such as that at Jabal Arkanu, does the metamorphic part of the basement seem somewhat elevated (Fig. 5.14.1). In such a case, it is overlain by Eocambrian or younger sandstone. The Precambrian intrusive component of the basement attains the highest elevation as it is more resistant.

Fig. 5.14.1 ***Jebel Al'Oweinat Basement exposure (Jabal Arkanu), SE Libya.***

The al-'Oweinat basement is made up of metamorphic rocks that range from simple quartzo-feldspathic gneiss to biotite-hornblende-hypersthene gneiss; thus at this end, the suite reaches the granulite facies. Marble and mylonite occur as minor interleaved rock constituents. The igneous rock component includes coarse-grained diorite and related rocks as well as a serpentinized gabbroic-pyroxenitic. The structural fabric of the basement has the imprint of three phases of folding and two episodes of faulting. The folding intensity appears to have weakened southward, apparently in association with relatively lower regional-grade metamorphism, but with a few local exceptions. Faulting is of the strike-slip type and assumes varying directions, the most prominent of which are ENE-WSW, NE-SW, ESE-WNW, and NW-SE. These involve right- and left-lateral shear couples. The almost E-W trend appears to be the youngest and it is along this direction that the diabase has been emplaced.

In contrast to the largely peneplaned OBP, the Pan-African (600–500 MA) TBP offers ample rock exposure. Based on the rock assemblage and the style of deformation, the area contains four metamorphic belts that tend to be lower in grade toward the west. In general, the metamorphic Tibestian terrain grades from garnet-staurolite schist, hornblende-to diopside-gneiss, meta-basalt, and calc-silicate rocks on the east to metagreywacke and sericite-chlorite schist on the west. Granites and related rocks, in addition to minor aphanitic equivalents, occur across the province, but they are more notable to the west. Intermediate to mafic rocks (such as diorite and gabbro) occur in lesser volume than felsic rocks. The intrusive rock component of the basement has been emplaced both syn-and post-tectonically. Cross cutting by quartz veins and dykes is not uncommon. Locally, some intrusions occur as simple ring structures in circular or oval forms, which can be observed in satellite images. The entire metamorphic terrain of the TBP carries the imprint of at least two phases of folding, but deformation is apparently more intense on the eastern side. Faulting across the province assumes two major trends of northeast–southwest and north-west–southeast. North–south faulting appears subordinate. Precambrian cataclasis is prominent in western Tibesti, especially in the Aouzou Belt where a quartz-injected, brecciated basement of chlorite schist is overlain by a massive sequence of Eocambrian to Cretaceous sandstone.

The Tibestian Basement in Libya disappears northward under a thick blanket of sand known as Sarir Tibesti. Sand and scattered outcrops of Paleozoic rocks limit basement exposure to the east and west. On the Chadian side, most of the basement disappears under the volcanic spewing of famed volcanoes such as Tarso Voon, Tarso Toon, and Pic Tousside. The GBP is dominated by granitic to granodioritic batholiths and minor diorite and gabbro of the Upper Proterozoic age that pierce through a metamorphic sequence of which only rafts are exposed in the southern part of the province. These rafts consist, in order of decreasing abundance (and from east to west), of phyllite and schist, granulite, and marble streaked with micaceous bands. The available information about uranium in Libya indicates the presence of uranium occurrence in the Ghat area in the southwestern part of the country. Uranium occurs within continental Permo-Triassic sandstone and siltstone along the flanks of the Murzuq basin (Assaf, and Aburkes, 1980), and the reported uranium minerals include tyuyammunite, carnotite, and betauranophane (Baegi et al., 1991).

Historical review of uranium resources

There is little information about the uranium resources in Libya. However, some scattered publications about radioactive anomalies and uranium mineralization exist. The regional geological and tectonic features of Libya have been reviewed in light of recognition criteria for the identification of uranium provinces. Three major basins are considered as potential uranium provinces. These basins are the Murzuq Basin, the Sarir Tibesti Trough, and the al-Kufra basin. It is noteworthy to mention that the granitoid rocks in southern Libya represent a potential uranium province (Salman, 1995).

The Libyan Atomic Energy Authority was established in 1973, and a department for the prospecting and exploration of uranium was established within this institution in 1974. The aim of this department is to study and identify potential areas for the presence of radioactive anomalies and to develop plans and programs for the discovery of uranium resources by the training of qualified technical personnel (Parslow and Karyma, 1995). According to geological considerations (geological structures, historical geology, orogeny, the ancient flow of water, and the presence of uranium source rocks and other radioactive materials) a region of about 250,000 square kilometers has been selected for exploration. Within this region, five areas were selected for the exploration of radioactive materials. These areas are the West al-'Owainat, the Etbiste, the Eastern region of al-'Owainat, the South Ghadames, and the Sirte Basin (Salman, 2005).

Some airborne reconnaissance surveys have been conducted in several areas including: al-'Owainat—Ghat, the South Ghadames, the Sirte Basin, and the northern Heroj. There has also been a ground-based follow-up for all detected radioactive anomalies, in addition to the detailed studies of radioactive anomalies for West al-'Owainat. Based on the geological criteria, the West al-'Owainat area represents the most promising area for uranium deposits. It includes sedimentary rocks of three consecutive times. These are lower Carboniferous, middle Carboniferous, upper Carboniferous, and Permian.

Detailed exploration work has been conducted, including emanometric, electric, geochemistry and drilling. Uranium mineralization was found on the surface and distributed within 44 radioactive anomalies with a content range from 114 to 4882 ppm. Moreover, uranium mineralization was found in subsurface, divided into three levels. Some secondary uranium minerals including tyamounit, carnotite, and beta uranophane were also identified (Baegi et al., 1991 and El Makhrouf et al., 1993). Based on the above, Libya has good potential uranium resources (Fig. 5.14.2).

Potential uranium resources

Based on a review of the available data for uranium resources and the general geology in Libya, some potential uranium resources can be suggested for further exploration.

Pan-African younger granites

The Bin Ghanimah batholith and the Tibesti massif are formed of Precambrian basement rocks and include some alkali-rich granites of the Pan-African age (Ghuma and Rogers, 1978). It is evident that major uranium deposits are associated with igneous activity in at least two Pan-African belts. The Rossing deposits of Namibia occur in pegmatite-alaskite migmatite bodies in the Damaran belt and broad areas of uranium mineralization are associated with emplacement of the younger granites of Egypt (Rogers, 1978). This fact can be utilized as a guide for uranium exploration in Pan-African granites and the associated igneous rocks in the Tibesti Mountains in Libya (Fig. 5.14.1). Moreover, these granites, acidic, and alkali volcanics can be considered also as source rocks for uranium.

Paleozoic-Mesozoic sedimentary sequences

These sediments show a strong potential for hosting economically viable uranium resources. This sequence is included in three giant basins in the southern region of Libya.

Murzuq basin

This basin lies in the southwestern part of Libya (Fig. 5.14.2) and extends about 700 km with a curvilinear axial trend following northeast in general and has about 400 km width. The basement is about 3000 m b.s.l. near the center of the basin (Goudarzi, 1978). The basin is filled with a thick sequence of clastic sediments formed mainly of sandstone and shales. These sediments are mainly continental and belong to the Paleozoic and Mesozoic ages.

This basin shows a strong potential for the presence of uranium deposits, with some uranium occurrences at its western flank in the Ghat area. The most interesting observation is its presence to the north of Niger and to the east of the Hoggar deposits (Wilpolt and Simov, 1979). One of the most interesting uranium discoveries in the past decade is that in the Mesozoic sandstones near Arlit in the Niger where reserves are estimated by the CEA to be at least 25,000 tons U30g (Bowie, 1970). Sandstone-type deposits west of the Air massif in Niger, on the eastern side of the Agades, production and reserves from carboniferous, Permian and Cretaceous sandstones, are recognized. Moreover, in the Lower Paleozoic Ajjer formation, Tassili series, Algeria, significant sandstone-type deposits are recognized, in addition to the presence of a vein and stockwork environment as a significant deposit in Late Proterozoic gneisses in the Hoggar massif to the southwest of this basin (Goodell, 1991). The Murzuq basin is surrounded by the same Carboniferous and Permo-Triassic units that have proved productive in nearby Niger; and the Djado Plateau of the Niger portion of the Murzuq basin has been prospected with some success.

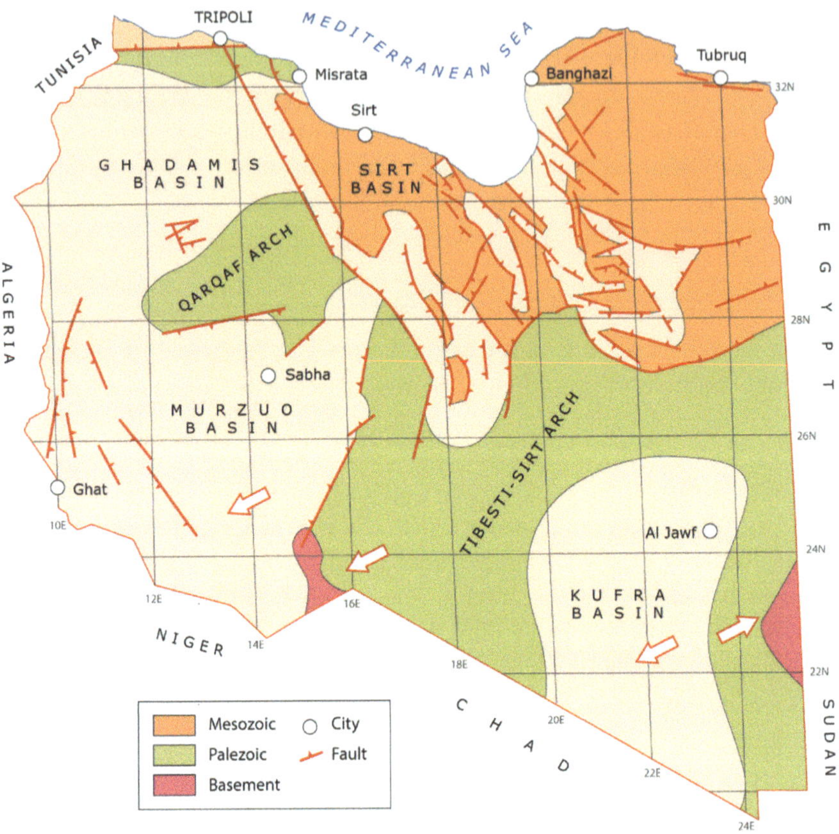

Fig. 5.14.2 ***Tectonic map of Libya and Egypt showing the major basins and some potential uranium areas (Arrows) (Modified after Salman, 2005).***

Sarir Tibesti Trough

The Sarir Tibesti Trough lies between Tibesti and the al-Haruj uplift in the west, Gebel Nuqay in the east, and the Tibesti Mountains in the south. The depth to the base in this trough is about 4000 m. Little information is known about the stratigraphy of this trough. However, some surface exposures of the Ordovician-Cambrian and Eocene are present. This basin is surrounded by basement rocks that include Pan-African younger granites in a "U-shape." Moreover, significant volcanic activity is noticed in the eastern and northern parts of this basin. This trough has received water and sediments from nearby source areas that are probably uraniferous. The area appears particularly interesting with respect to uranium potential (Salman, 1995).

Al-Kufrah basin

This basin is in the southeastern part of Libya with an axial trace following NNE. It is limited from the east by the Gebel al-'Uwaynat uplift, the Gebel Az Zalamah uplift from the northwest, and the Tibesti mountains from the southeast (Goudarzi, 1978). The basin is covered with a thick sequence of sedimentary rocks of the Paleozoic and Mesozoic ages. The Paleozoic sediments range in thickness from 1055 to 1850 m, and its lower part (1300 m) is continental, mainly fluviatile and locally marine. The upper part (550 m) shows oscillation from continental to marine. The Mesozoic sediments ranges in thickness from 200 m to 735 m and they are mainly continental (Bellini and 1978). This basin is a good target for uranium exploration, based on the presence of a thick succession of Paleozoic and Mesozoic mainly continental clastic sediments. Moreover, some radioactive anomalies are recorded in the basal conglomerates and Paleozoic sandstones in Egypt near Gebel al-'Uwaynat (El Aassy, 1984).

CHAPTER 5.15

Morocco

Geological outlines

Five major geological domains are distinguished in Morocco: (1) The Anti Atlas and Saharan domain, comprising gently deformed Paleozoic sealed to the Northwest African Craton during the Hercynian orogeny; (2) the Mesetian domain, characterized by deformed Paleozoic unconformably overlain by gently folded Meso-Cenozoic deposits; (3) the Atlasic Domain corresponding to the inversion of the Atlasic that was filled by thick Mesozoic sediments; (4) the Atlantic passive margin, consisting of thick Mesozoic sediments locally deformed by salt tectonics and gravity-induced imbrications with important turbidite sheet deposit basin wards; and (5) the Rift domain, a thrust fold belt where the deformation of the Meso-Cenozoic series decreases southward (Fig. 5.15.1).

Tectonically, Morocco was marked by two major compressional events (Hercynian and Alpine orogenies) separated by an extensional period related to the Atlantic opening. This represents the post-Precambrian structural evolution. The geology of Morocco was impacted by the Hercynian Orogeny, the Triassic-Liassic extension, and the Alpine Orogeny. The Upper

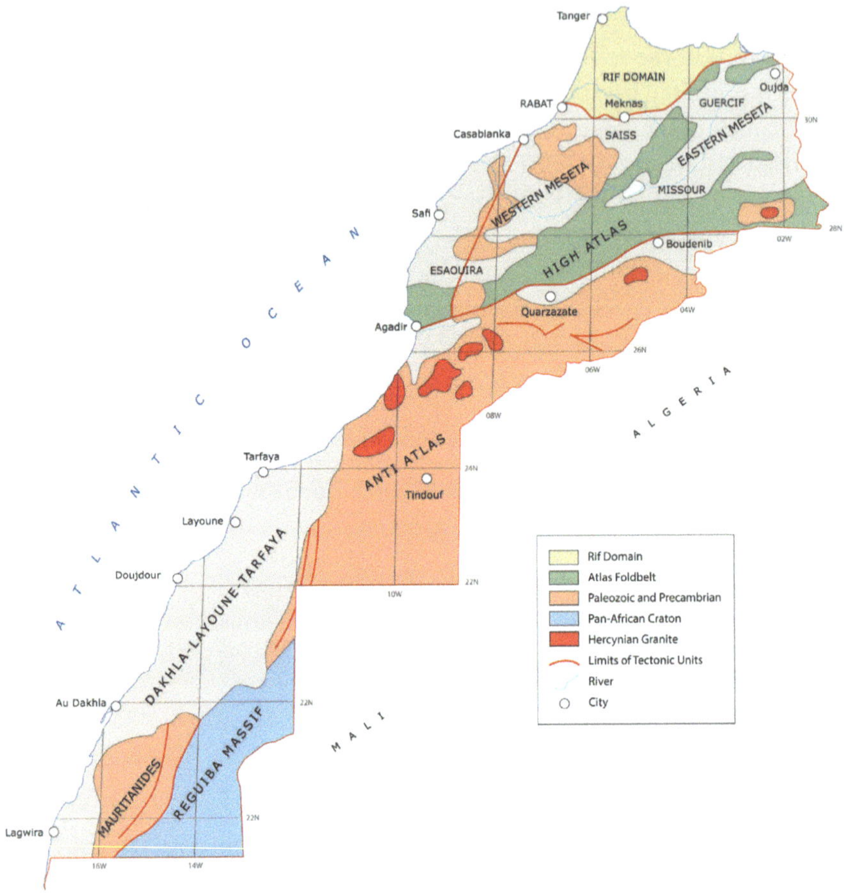

Fig. 5.15.1 ***General geological and tectonic map of Morocco.***

Devonian-Carboniferous compression resulted in folding and faulting of the Paleozoic series. The remnants of the hercynian fold belt crop out in the Anti Atlas and the Moroccan meseta. The opening of the Atlantic was preceded by the Triassic-Lower Jurassic rifting which was followed by massive regional subsidence during the Jurassic and much of the Lower Cretaceous in the Atlantic passive margin. This extension was also responsible for the opening of the Atlas troughs. During the Upper Eocene-Oligocene, the Atlas Mesozoic troughs were inverted to form the High and Middle Atlas. The late collision of Africa with Europe during the Neogene (Int. Mo.1) resulted in the formation of the Rift which is a segment of the Western Mediterranean Alpine fold belt.

Historical review of uranium exploration

There was an active mission from the French Atomic Energy Agency (CEA) for uranium exploration in Morocco in 1946–1947. This mission started cooperation with the Department of Mining in Morocco from 1951 to 1953. In 1953, a joint Moroccan French company (SOMAREM), with support from the United States, was founded for the purpose of conducting mining research. The activities of this company continued until 1956 and performed general mineral exploration in all of the Moroccan territories. The interest of this company was the vein-type deposits and later began to search for similar types in the American Colorado Plateau. They used airborne, car-borne, and foot exploration surveys. In the West Atlas Heights, a layer of sandstone uranium-bearing continental sandstone type of the Cretaceous Age was discovered in the Ayman Tanot area (Marrakech region). A program for core and percussion drilling for gas was proposed to cover an area of 15 km^2 in search of radioactive anomalies.

In 1970, a research program on metallic ores of Anti Atlas area was conducted by the United Nations and the Bureau de Recherches Géologiques et Minières (BRGM). A regional radiometric airborne survey was included, which covered 1800 km^2, in cooperation with the International Atomic Energy Agency. In 1973, as part of IAEA aid to Morocco, one of the BRGM experts summarized previous studies and proposed recommendations for the target area. Since 1974, the BRGM has made great efforts in the field of exploration, implementing numerous methods and forming integrated geological teams to train and equip personnel. A geological reconnaissance study on Bermo-Wervenyn in the Beni Madden area was carried out, proving the existence of uranium and giving home to the possible presence of larger uranium deposits in Morocco. According to the attained results, two projects were selected in Upper Moloya and West Atlas Heights for study. In Upper Moloya, a reconnaissance survey has been conducted. Initial results have proven the existence of many radioactive anomalies at the base of sedimentary rocks of Triassic age, and therefore, have been selected for the performance of exploratory core drilling to define the main geological features.

In 1979 and 1980, exploratory activity took place in the Cretaceous age rocks in the Huot Atlas Aksidntal, in the Permo-triassic rocks of the Moloya area, and in areas of Precambrian rocks in the Ante Atlas. A geochemical exploration program was also implemented for prospecting in the basement rocks at Nordmstan with the assistance of the International Atomic Energy

Agency (IAEA). In 1981 and 1982, exploration continued in the continental rocks of the Cretaceous age in the Atlas Mountains Heights (Marrakech Province). There has also been additional exploration in the Tertiary Basin in Ouarzazate, near the mid-Moloya River; in areas of Precambrian exposures in Anti Atlas an airborne radiometric survey has been conducted in the northeastern region of Marrakesh.

Between 1983 and 1984, a program for spectrometric surveys in the Hout Atlas Occidental was postponed because of budget concerns. This program was replaced by geological and radiometric surveys, as well as drilling in selected areas in the Anti Atlas. In 1985 a helicopter was used in a magnetic, electromagnetic, and spectrometric survey for uranium and base metals covering 8000 km^2 in the Atlas Heights, leading to the discovery of several radioactive anomalies for uranium and uranium–thorium. It is noteworthy to mention that 204 wells have been drilled with a total length of 25,300 m in the areas of exploration in Morocco until the end of 1984. The total expenditure on these exploratory operations in 1987 reached 2.25 million DH in addition to 2.752 billion USD. The results of these efforts were not encouraging, as the concentration of uranium was not economically viable. Accordingly, further works were not recommended.

Uranium resources

According to the country report of Morocco in the Red Book released by the International Atomic Energy Agency in 1989, there were no conventional uranium resources in Morocco (Uranium, 1989). However, the book indicated that there are many known uranium occurrences in the Anti Atlas, Hout Atlas, Occidental East, Mesta and Mowen Atlas, and Hout Moloya regions. From the geological point of view, these radioactive occurrences are located in the Precambrian and Paleozoic granites, as well as in the sedimentary units of the Cambrian and Cretaceous. The estimated reserves in 1983, which fall within the group of Speculative Resources, range between 70,000 to 180,000 tons of uranium.

Potential uranium resources

Based on the available data for uranium resources and the general geology of Morocco, the following environmental regions are potential targets for future exploration:

Pan-African granites

The reevaluation of sites is necessary to survey possible uranium occurrences in the areas of the Atlas Mountains, especially the double mica younger granites of the Pan-African type (younger granites), which are in the age range of 400–600 million years. The fault zones and intersections represent target sites for hosting uranium resources. Small- to medium-sized uranium deposits can be expected in these geological environments.

Intra-cratonic basins

Further attention should be paid to the intra-cratonic basins that include some low-grade metamorphosed sediments as seen in the Hammamat series in Egypt of the late Proterozoic age. These basins can form suitable traps for uranium resources, especially if there are some uranium-bearing granite or volcanic rocks in contact or nearby these basins.

Basal conglomerate and the Paleozoic clastics

The basal conglomerate that covers the Precambrian basement rocks can be considered as an important target for uranium exploration in Morocco. It is indicated from similar studies in the Eastern and Sinai Deserts, Egypt, that this geological environment includes some radioactive anomalies. Also, the clastic sedimentary sequence of the Paleozoic represents a good target for uranium exploration, forming a good trap for uranium resources. The Basin structures, which include these Paleozoic clastic sediments, are promising sites for uranium resources if the prevailing geochemical environments are suitable for uranium deposition. The uranium-bearing sandstone types are not uncommon in these structures. Where there are prospects for the discovery of raw materials for pregnant sandstone uranium type Sandstone as those basins are suitable for hosting uranium if there is a close source of rocks and if the geochemical environment is suitable for the deposition of uranium.

Phosphate deposits

Phosphate deposits in Morocco have the highest amount of uranium content relative to other phosphate deposits in the Middle East and North Africa (Tables 5.15.1 and 5.15.2). However, the direct extraction of uranium from phosphate ores is generally not economically viable. Uranium can be extracted as a byproduct from phosphoric acid, and therefore the unit for uranium extraction should be built in the same location of the

Table 5.15.1 Reserves of uranium in phosphate ores in Morocco.

Deposit name	Site	Ore tonnage	Grade (g/t)
Awlad Abdoun	**Khurijba**	3,220,000	120
Jantor	**Youssovia**	966,000	130
	Ben Jorrier	2,40,000	180
Aud El Dahab	**Youkra**	57,000	60
Mscala	**Asoeria**	2,043,000	100

Table 5.15.2 The main characteristics of uranium resources in Morocco.

Uranium deposits	Characteristics
Unconventional	The phosphate deposits in Morocco are huge in amount and reach 56,250 million tons with 28% as average P_2O_5 (Prévot, 1990). They have the highest amount of uranium content relative to other phosphate deposits in the Middle East and North Africa. The uranium content in these phosphorites ranges from 60 to 180 ppm (Salman, 2005). The estimated uranium content is 6.9 million tons (IAEA, 2001), (Type localities: Khurijba, Youssovia, Ben Jorrier, Youkra and Asoeria).

phosphoric acid plant. It should be noted that this unit will be economically viable, as in addition to uranium extraction it will be used in the purification of sulfuric acid where some harmful elements will be disposed of. A high degree of purity makes the phosphoric acid viable for industrial, medical, and agricultural uses, raising the economic visibility of the unit.

CHAPTER 5.16

Tunisia

Geological outlines

Geologically, Tunisia represents a limited part of the foreland of the African continent. The Dorsal, an extension of the Atlas Mountains, traverses Tunisia in a northeasterly direction from the Algerian border in the west to the Cape Bon peninsula. North of the Dorsal is the Tell, a region characterized by low, rolling hills and plains, but reaches elevations of 1050 m in the northwestern corner of Tunisia. The Sahil is a plain along Tunisia's eastern Mediterranean coast famous for its olive monoculture. Inland from the Sahil, between the Dorsal and a range of hills south of Gafsa, much of the southern region is semi-arid and desert.

The main rock exposures belong to the Phanerozoic sedimentary sequences. In the southern half of Tunisia, the Triassic, Jurassic, and lower Cretaceous are exposed in the Eastern parts. The upper Cretaceous exposures form a huge belt following the NW-SE direction and cover most of the central zone of the southern part of the country (Fig. 5.16.1). The northern half is geologically complicated, due to the effect of numerous faults that form several horsts, grabens, and fault terraces, acting almost as a continuation of the Atlas Mountains. The Palaeozoic exposures include units from Permian to Pliocene–Miocene and quaternary. Here, the Cainozoic volcanics are not uncommon (Fig. 5.16.1).

Uranium resources

No reports about uranium resources were included in the IAEA uranium ledgers. However, Tunisia's phosphates, base metals, and petroleum industry provide most of its foreign earnings. Phosphates alone accounted for nearly 45% of the country's export earnings.

Potential uranium resources

The type and mode of distribution and of geological rock unit exposures often reflect the potential for uranium resources. Also, the presence of several structural features especially horst, grabens, and shear zones can represent a suitable trap for uranium localization. Moreover, the volcanic activities during the Cainozoic Period can represent a catalyst agent for redistribution or as source rocks if they are of the acidic or alkaline types.

Phosphate deposits

The phosphate deposits in Tunisia are exclusively sedimentary and belong to the Tethyan Phosphate belt, which are of the Upper Cretaceous age. Tunisia's deposits are also very rich in the carbonate-fluorapatite form of phosphate, otherwise known as francolite. The average uranium content in Tunisian phosphate is 75 ppm. There is a strong potential to find uranium resources within these phosphate beds. Moreover, some secondary uranium mineralization could be detected as fissures and vein filling in Tunisian

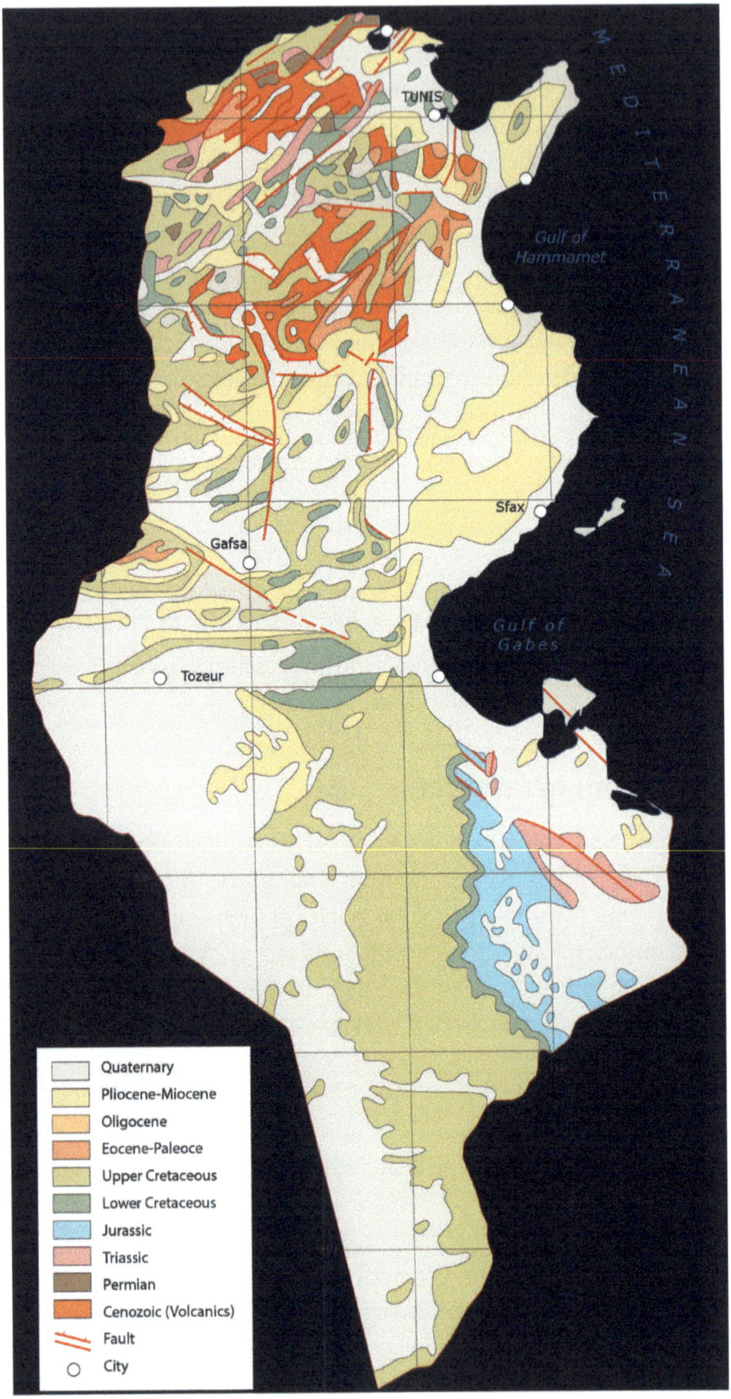

Fig. 5.16.1 ***Geological map of Tunisia.***

phosphate horizons as recorded in Syria and Jordan. Also, if there are black shale horizons associated with the phosphate-bearing sediments in Tunisia, then there is a good chance that they also hold uranium deposits.

Volcanic rocks

Volcanic rocks are exposed in Tunisia in several locations (Fig. 5.16.1). The acidic and alkaline types can be considered as a source for uranium mineralization. Moreover, the basic volcanic dykes can act as barriers for trapping uranium. The geological and structure features at the site can play an important role in the localization of uranium resources in some of these volcanics.

Recent uranium deposits

There is a strong potential for the localization of recent uranium deposits in Tunisia. The most promising sites are those near the phosphate-bearing sediments. The basins and grabens structures in the northern region of the country, which represent the continuation of the Atlas Mountains, are very suitable for hosting this type of uranium resources. Moreover, these structures within the NW-SE trending Upper Cretaceous sediments (Fig. 5.16.1), which bear phosphate deposits, are very promising for hosting recent uranium deposits.

References

Abdel Meguid, A.A., 1986. Geologic and radiometric studies of uraniferous granite in Urn Ara–Urn Shilman Area, South Eastern Desert, Egypt. (Ph.D. thesis), Fac. Sci. Suez Canal University, Egypt, p. 241.

Abdel Monem, A., Bakhit, F., Ali, M.M., 1990. Trace and rare earth elements geochemistry of the uranium mineralization at El-Erediya, central Eastern Desert, Egypt. Egypt. Mineral. 2, 143–150.

Abdel-Monem, A.A., Hurley, P., 1979. Evolution and Mineralization of the Arabian-Nubian Shield: Proceedings of a Symposium, 3. Institute of Applied Geology, King Abdulaziz University, p. 141.

Abou Deif, A., 1985. Uranium mineralization along Qena-Safaga road. (M.Sc. thesis). AI Azhar University, Cairo, p. 109.

Abu Deif, A., 1992. The relation between the uranium mineralization and tectonics in some Pan-African granites, west of Safaga, Eastern Desert, Egypt. (Ph.D.thesis). Assiut University, Egypt, p. 218.

Afifi Sofia, Y., 1991. Geochemical investigation of some radioactive lower Carboniferous sedimentary rocks of Wadi Nasib area, West Central Sinai, Egypt: (M.Sc, thesis). Suez Canal Univ., Egypt, p. 250.

Al-Bassam, A.S., Al-Dahan, A.A., Jamil, A.K., 1983. Campanian-Maastrichtian Phosphorites of IraqMiner. Deposita.18, 215–233.

Alfotawi, B.A., Gazzaz, M.A., Hassan, A., 1991. Uranium and thorium environment in Haql granite in Midyan Region, Northwest of Saudi Arabia. Earth Sci. J. 4, 161–170.

Al-Laboun, A.A., 1987. Distribution of Permo-Carboniferous clastics of Greater Arabian basin. AAPG (Am. Assoc. Pet. Geol.) Bull.; (United States) 71, no. CONF-870606-.
Al-Saleh, A.M., Boyle, A.P., Mussett, A.E., 1998. Metamorphism and 40Ar/39Ar dating of the Halaban ophiolite and associated units: evidence for two-stage orogenesis in the eastern Arabian shield. J. Geol. Soc. 155 (1), 165–175.
Almehdi, B., Gojkovic, S., Megrisi, M., obrenovic, tvt, Puric, D., Zelenka, J., 1991. Radioactive elements in sedimentary rocks of the western part of Murzuq Basin, Proc. Third Symposium on the Geology of Libya, held in Tripoli. Libya, September, 27-30, 1987, V. VII, 2645–2658.
Anace, S., N. Birtek, N., Birsen, N., Kafadar, M.S., 1985. Exploration of uranium resources in Turkey: A short review in retrospect of the technical and economical aspects. p. 10.
Assaf, H.S., Aburkes, M.G., 1980. Uranium occurrence in Ghat area, Southwest Libya. Ann. Geo. Surv. Egypt V (X), 871–879.
Attawiya, M.Y., 1983. Mineralogical study of E1 Erediya-l uranium occurrence, Eastern Desert, Egypt. Arab J. Nucl. Sci. Appl. 16 (2), 221–235.
Baegi, M.B., Assaf, H.S., Hangari, K.M., 1991. Al Awaynat surface uranium mineralization—A new approach to its origin, Proc. Third Symposium on the Geology of Libya, held at Tripoli, September 27-30, 1987, V. VII, 2619–2626.
Bakhit, F.S., El Kassas, I.A., 1989. Distribution and orientation of radioactive veins in E1 Erediya-EI Missikat area, central Eastern Desert, Egypt. Int. J. Remote Sens. 10, 565–581.
Bakhit, F.S., Guirguis, L.A., 1983. Contribution to the granite plutons of both EI Missikat and El Erediya, Central Eastern Desert, Egypt. Communication. Dergi Yayin Skreterligi, Ankara University, Fen Fakultesi. Besevler- Ankara;-Turkey, Ser, C1, 1–7.
Baturin, G.N., Kochenov, A.V., 2001. Uranium in phosphoritesLithol. Miner. Resour. 36 (4), 303–321.
Bellini, E., Massa, D., 1978. A stratigraphic contribution to the Paleozoic of the southern basins of Libya, Proc.Second Symposium on the Geology of Libya. Tripoli, V. I, 3–56.
Ben Atiq, M.A., 2005. Origin of manganese mineralization in the sediments of shammar group in the Asfar Thwelil Area, Northern Arabian Shield (M.Sc.Thesis). Department of Geology, Fac, Sc. King Saud University.
Bergougnan, H., 1975. Dispositif des ophiolites nord-est anatoliennes, origine des nappes ophiolitiques et sud-pontiques, jeu de la faille nord-anatolienne. C. R. Hebd. Seances Acad. Sci. 281, 107–110.
Bowie, S.H.D., 1970. World uranium deposits. In: Proc. Uranium Exploration Geology Panel, Vienna, 13-17 April, 1970 (IAEA-PL-391/19), pp. 23–34.
Bozkurt, E., Satir, M., 2000. The southern Menderes Massif (western Turkey); geochronology and exhumation history. Geol. J. 35, 285–296.
Brew, G., Barazangi, M., Sawaf, T., Al-Maleh, K., 2000. Tectonic map and geologic evolution of Syria: the role of GIS. Lead. Edge 19, 176–182.
Cameron, J., 1963. Report to IAEA, Development and Evaluation of Uranium Deposits. Oct. 1962 Sep. 1963.
Chaimov, T., Barazangi, M., Al-Saad, D., Sawaf, T., 1990. Crustal shortening in the Palmyride fold belt, Syria, and implications for movement along the Dead Sea fault system. Tectonics, 9 (6), 1369-1386.
Dabbour, G.A., 1994. The Egyptian placer deposits—a potential source for nuclear raw materials, Proc. Second Arab Conference on the Peaceful Use of Atomic Energy. Cairo, Egypt, 5-9 November, 1994. pp. 191–204.
Dahlkamp, FJ., 1987. Classification scheme for uranium ore deposits. A state of the art review, Proc. Technical Committee. Meeting on Meta~logenesis of Uranium Deposits organized by The IAEA. Vienna, p. 31.
De Voto, R.H., 1982. Uranium Exploration. In: deVivo, B (Ed.), Uranium Geochemistry, Mineralogy, Geology, Exploration and Resources. Springer, Amsterdam, pp. 101–108.

De Voto, R.H., Stevens, D.N., 1979. Uraniferous phosphate resources and technology and economics of uranium recovery from phosphate resources, United States and Free World (No. GJBX–110 (79)(VOL. 1)). Earth Sciences.
Duyverman, H.J., 1984. Late Precambrian granitic and volcanic rock and their rule to the cratonisation of the Arabian Shield, Proc. Symposium on Pan-African Crustal Evolution in the Arabian –Nubian Shield. G.C.P. No 164, 1-3 Feb. 1982, K.A. Univ., F.E.S. BULL., 6, 49–69.
Evans, M.I. (Ed.), Important Bird Areas in the Middle East. BirdLife Conservation Series 2. BirdLife International, Cambridge, U.K. p. 410.
Eid, M.A., Fakhry, A.A., Mahdy, A.A., Eid, K.A., Broekaert, J.A.C., 1995. A contribution to the evaluation of REE in an Egyptian phosphate deposit (Abstract): Fresenius. J. Anal. Chem. 351 (2-3), 190–196.
El Assay, E.E., 1984. Contribution to the radioactivity of Palaeozoic rocks in Gebel Oweinat Area. South-western Desert, Egypt: Arab. J. Nuclear Sci. App. 17 (2), 339–355.
El Khouly, D.M., 1996. Geology and Radioactivity of Esh El Mellaha Range, with emphasis on the granites of G. Homera El Girigab, north Eastern Desert, Egypt (Ph.D. Thesis). Cairo University, Cairo, Egypt.
El Ramly, M.F., Hussein, A.A.A., 1985. The ring complexes of the Eastern Desert of Egypt. J. Afr. Earth Sci. (1983) 3 (1-2), 77–82.
El Shatoury, H.M., Mostafa, E.M., Nasser, F.S., 1984. Granites and granitoid rocks in Egypt. A statistical approach of classification. Chem. Erde. 43, 229–246.
El-Kammar, A.M., Ali, B.H., El-Badry, A.M., 2009. Environmental geochemistry of river Nile bottom sediments between Aswan and Isna, Upper Egypt. J. Appl. Sci. Res. 5 (6), 585–594.
El-Kammar, A.M., Salman, A.E., Mahdy, A.I., 2001. Geochemical and genetical constraints on rare metals mineralization at the central Eastern Desert of Egypt. Geochem. J. 35 (2), 117–135.
ELKattan, E.M., El Aassy, I.E., 1992. Structurally controlled radioactivity in the Um-Bogma Area, Midwestern-S'inai, Egypt: the geochemistry and uranium favourablity of the postorogenic granites of the Northwestern Arabian Shield, Kingdom of Saudi Arabia. Bull. Fac. Earth Sci, King Abdulaziz Univ. 6, 195–209.
Gaby, EI, S., K.F.List, Therani, R., 1990. The basement complex of the Eastern Desert and Sinai Egypt. In: Said, R., Balkema, A.A. (Eds.), The Geology of Egypt. Egyptian General Petroleum Corp, Rotterdam, Netherland, pp. 175–184.
Ghanem, M., Mikhailove, L.A., Zalata, A.A., Razvaliacv, A.V., Abdel Razik, I.M., Nirtove, Y.V., 1970. Stratigraphy of the phosphate bearing Cretaceous and Palaeogene sediments of the Nile Valley between Idfu and Qena. Extract from O. Moharram et al. Studies on some mineral deposits of Egypt, Part II, Article 7. Geol. Surv. Egypt 109–131.
Ghuma, M.A., Rogers, J.W., 1978. Pan African evolution in J amahiria and North Africa, Proc. Second Symposium on the Geology of Libya. Tripoli, V, 1059–1064.
Goodell, P.C., Waters, A.C., 1981. Uranium in volcanic and volcaniclastic rocks. American Association of Petroleum Geologists, OK, Volume 13.
Goodell, P.C., 1991. Uranium potential in and around the Tibisti and Hoggar Massifs, Proc. Third Symposium on the Geology of Libya. Tripoli, September, 27-30, 1987, V. VII, 2627–2636.
Goudarzi, G.H., 1978. Structure—Libya: In: Proc. Second Symposium on the Geology of Libya. Tripoli, V, 879–892 September 16-21, 1978.
Greenberg, J.K., 1981. Characteristics and origin of Egyptian younger granitess. Geol. Soc. Am. Bull. part II, 92, 749–840.
Huffnagel 11, 1980. Investigation of the El-Lajjun Oil Shale Deposit. Internal unpublished report. Natural Resource Authority, Amman, Jordan.
Hamarneh, Y., 1998. Oil shale resources development in Jordan. Revised and updated by. Alali, Jamal, Sawaqed, Suzan (Eds.), November 2006. Natural Resource Authority, Amman, Jordan, p. 101.

Hamdi, B., Brasier, M.D., Zhiwen, J., 1989. Earliest skeletal fossils from Precambrian–Cambrian boundary strata, Elburz Mountains, Iran. Geol. Mag. 126 (3), 283–289.
Hashad, A.H., Hassa, M.A., Abu El Gadayel, A.A., 1982. Geological and petrographical study of Wadi Natash, Late Cretaceous volcanics. Egypt J. Geol. 26, 19–37.
Hassan, M.A., 1964. Geology and petrographical studies of the radioactive minerals and rocks in Wadi Sikait-Wadi El-Gemal area Eastern Desert. UARM Sc thesis, Faculty of Science, Cairo University, Chicago.
Hassan, M.A., 1973. Geology and geochemistry of radioactive columbite-bearing psammitic gneiss of Wadi Abu Rusheid. South Eastern Desert, Egypt [J]. Annal. Geol. Surv. 3, 207.
Hussein, A.H., Hassan, M.A., El Taher, M.A., Abou Deif, A., 1986. Uranium bearing silicious veins in younger granites, Eastern Desert, Egypt. IAEA, TECDOC IAEA, Vienna, 361, pp. 143–157.
Hussein, H.A.M., Sayyah, T.A., 1992. Uranium potential of the younger granites of Egypt: Report no. IAEA- TECDOC-650. IAEA, Vienna, pp. 58–64.
IAEA, 1992. Report on the 12th IAEA On-Site Inspection in Iraq under Security Council Resolution 687, 2 July 1992, UN Doc S/24223.
IAEA, 1993. Report on the 19th IAEA On-Site Inspection in Iraq under Security Council Resolution 687, 21 June 1993, UN Doc S/25982.
IAEA, 2001. https://www.iaea.org/sites/default/files/anrep2001_full.pdf.
IUCNIUNEP, 1985. The management and conservation of renewable marine resources in the Kuwait Action Plan region. UNEP Regional Seas Reports and Studies 56, 57.
Ibrahim Salman, A.B., 1968. Geology and radioactivity of Wadi El Kareem- Wadi El Oweirsha area, Eastern Desert, Egypt (M.Sc. Thesis). Ain Shams Univ., Cairo, Egypt.
Ibrahim, M.E., Saleh, G.M., El-Naby, Abd, H.H., 2001. Uranium mineralization in the two mica granite of Gabal Ribdab area, South Eastern Desert, Egypt. Appl. Radiat. Isot. 55 (6), 861–872.
Ibrahim, M.E., Zalata, A.A., Hasssaf, H.S., Ibrahim, L.H., Rashed, M.A., 2003. El Sella shear zone, South Eastern Desert, Egypt: an example of veiun type uranium deposit. Egypt. J. Geol. 47 (2), 689–704.
Jami, Morteza, 2005. Geology, Geochemistry and Evolution of the Esfordi Phosphate—Iron Deposit. University of New South Wales, Bafq Area, Central Iran. http://www.library.unsw.edu.au/~thesis/adt-UN/public/adt-NUN20060905.145823/index.html.
Johnson, P.R., 1998. Tectonic Map of Saudi Arabia and Adjacent Areas (scale). Deputy Ministry for Mineral Resources, Jiddah, Saudi Arabia.
Klitzsch, E., 1990. Paleozoic: in The Geology of Egypt, R. Said, ed. ch. 21 A.A. Balkema, Rotterdam, Netherland, 393-406. Radioelements distribution in some granitoid rocks from the North Eastern Desert of Egypt. Mans. Sci. Bull. 16 (1), 17–39.
Master, S., Woldai, T., 2004. THa Umm Al Benni structure, in the Mesopotamian Marsh lands of south Iraq, as a postulated Late Holocene Meteorite impact crater: Geological setting and new Landsat ETM+ and Aster Satellite imagery. University of the Witwatersrand, Johannesburg, Economic Geology Research Institute, p. 21. circular No.
MTA, 1978. Manisa-K.prübasi Uranyum Cevherleri Uzerine Pilot Capta Teknolojik Calisma. Aralik 1978.
Nagati, M., 1986. Possible Mesozoic rifts in Upper Egypt: An anology with the geology of Yemen-Somalia rift basins., Proc. EGPC 8th Exploration Conference. Cairo, Egypt, p. 20.
NEDDD, 1998. Internal report about exploration activities in J. Qattar, NMA, Cairo, Egypt, p. 25.
Nardulli, Giuseppe, 1998. Nuclear Weapons in the Middle East, Seminar held at the Summer School On the Diplomatic Aspect of the Middle East Peace Process. Molfetta, Italy, 7-17 September 1998.
Noweir, A.M., Lebda, E.M., Nassim, A.S., El Kholeify, E.M., 2003. Tectonic evolution, geochemistry and petrogenesis of the alkaline volcanic rocks of Wadi Natash. Egypt. J. Geol 47 (2), 541–568.
Olexcon Int., 1967. Micropalaeontology, palynology, stragraphy and oil source rock identification; various wells and sections. Unpublished Report, Nat. Res. Auth. Amman.

Osman, A.M., Salman, A.B., Atawiya, M.Y., Khalaf, E.A., Khalaf, M., 2001. Metallogenesis of uraniferous subsurface younger granites, Um Ara Area, South Eastern Desert, Egypt., Proc. Second International Conference on the Geology of Africa, V.II, Assiut University. Egypt, pp. 199–218.

Parslow, G.R., Karyma, R., 1995. Economic potential of black shale: The Bakken formation in Saskatchewan (No. IAEA-TECDOC–823).

Parslow, G.R., Khalil, B., Hassan, A.R.K., 1992. Baseline uranium exploration surveys and their use in the early detection of potential natural hazards: a case history from Sudan, Proc. Technical Committee Meeting. IAEA- TECDOC-650, Vienna, pp. 22–29.

Powers, R.W., Ramirez, L.F., D, C., Redmond, Elberg Jr., E.L., 1966. Geology of the Arabian Peninsula, sedimentary geology of Saudi Arabia. U.S. Geological Survey Professional Paper 560-D, 147.

Ponikarov, V.P., 1966. The Geology of Syria. Explanatory Notes on the Geological Map of Syria, Scale 1:200 000. Ministry of Industry, Damascus, Syrian Arab Republic.

Prévot, M., Derder, M.E.M., McWilliams, M., Thompson, J., 1990. Intensity of the Earth's magnetic field: evidence for a Mesozoic dipole low. Earth Planet. Sci. Lett. 97 (1-2), 129–139.

Rahman Abdel, I.F., Gh. Mansour, M., 2005. Microbial control on the mineralogy of Arief El Naqa phosphates, Egypt (Abstract): Acta Geol. Hung. 48 (3), 235–350.

Rice, S.P., Robertson, A.H.F., Ustaömer, T., 2006. Late Cretaceous-Early Cenozoic tectonic evolution of the Eurasian active margin in the Central and Eastern Pontides, northern Turkey. In: Robertson, Alastair H.F. (Ed.). Tectonic Development of the Eastern Mediterranean Region. Geological Society, London, Special Publications, London, 260, pp. 413–445.

Riddler, G.P., Eck, M.V., Aspinall, N.C., McHugh, J.J., Parker, T.W.H., Farasani, A.M., Dini, S.M., 1986. Saudi Arabian deputy ministry for mineral resources. Jeddah, technical record RF-TR-06-2.

Robertson, A., Dixon, J.E.D., 1984. Introduction: aspects of the geological evolution of the Eastern Mediterranean. In: Dixon, J.E.D., Robertson, A. (Eds.). The Geological Evolution of the Eastern Mediterranean. Geological Society, London, Special Publications, 17, pp. 1–74.

Rogers, J.W., 1978, Uranium in Pan-African Belts: In: Proc. Second Symposium on the Geology of Libya. Tripoli, Libya, V. III, 1045–1049.

Sabet, A., Tsogoev, V., Baburin, L., Raid, A., Zakhari, A., Armanius, L., 1976. Geologic structure and laws of localization of tantalummineral zonation at the Nuweibi deposit. Ann. Geol. Surv. Egypt VI, 119–156.

Sadig, A., Ahmed, Y.M., Charbonneau, B.W., Lecheminant, G.M., 1988. The Uro radiometric anomaly—uraniferous phosphate in a tectonic breccia (Nuba tvlountains), Kordofan Province, Sudan Uranium 4 (24), 351–363.

Saeidi, Mohamed, 2005. Nuclear fuel cycle activities in Iran, Proc. World Nuclear Association Annual Symposium 7-9 September 2005. London, pp. 1–10.

Said, R., 1962. The Geology of Egypt. Elsevier Publisher. Co., Amsterdam, New York.

Salman, A.B., 2008. Notes and observations during the field visit to central Jordan recent uranium deposits, Proc. IAEA Technical Meeting on 'Uranium Exploration and Mining Methods', November 17-20, 2008, Amman, Jordan.

Salman, A.B., 2005. Nuclear Age: Uranium Exploration and its Reserve Estimation and Potential Provinces in Some Arabian Countries. Dar El Ketab Al Hadeth, Cairo, Egypt, p. 295. ISBN: 977-350-105-1 (In Arabic).

Salman, A.B., 1995. Potential uranium provinces in some Arabian countries, Proc. Second Arab Conference on the peaceful uses of atomic energy, Cairo, Egypt, 5-9/11/1994. Tunis). Arab Atomic Energy Agency, pp. 205–227.

Salman, A.B., Shalaby, M.H., Nossair, L.M., 1990. Uranium province, North Red Sea Hills, Egypt, Proc. International Earth Sciences Congress on Aegean Regions. Izmir, Turkey, 1-6 October 1990, 1, 89–101.

Salman, A.B., 1974. Structure and Radioactivity of Some Phosphate Deposits, East Luxor Area, Eastern Desert, Egypt (Ph.D. Thesis). Ain Shams University, Cairo, Egypt, p. 322.

Salman, A.B., Aassy, I.E., 1983. Radioactivity and uranium distribution in Wadi Araba Late Paleozoic sediments, north Eastern Desert, Egypt. Ann. Geo. Surv. Egypt 13, 53–60.
Salman, A.B., 1983. Uranium potentiality and significance of the bostonite generations, central Eastern Desert, Egypt. Ain Shams Med. J. 28 (1982-1983), 63–81.
Salman, A.B., El Aassy, I.E., Shalaby, M.H., 1990. New occurrence of uranium mineralization in Gabal Qattar, Northern Eastern Desert, Egypt. Ann. Geo. Surv. Egypt V. XVI (1986-1989), 31–34.
Salman, A.B., Sheshtawi, Y.A., Aly, M.M., Hamoudah, E.A., 1989. Radioelements distribution in some granitoid rocks from the North Eastern Desert of Egypt. Mans. Sci. Bull. 16 (1), 17–39.
Sawaf, T., Al-Saad, D., Gebran, A., Barazangi, M., Best, J.A., Chaimov, T., 1993. Structure and stratigraphy of eastern Syria across the Euphrates depression. Tectonophysics 220, 267–281.
Sayyah, T.A., Attawiya, M.Y., 1989. Contribution to the mineralogy of uranium occurrence of G. Qattar, Eastern Desert. Egypt. Arab J. Nucl. Sci. App. 23 (1), 117–184.
Spanderashvilli, G.I., Mansour, M., 1970. The Egyptian phosphates others. In: Moharram, O. (Ed.), Studies on Some Mineral Deposits of Egypt. UAR Geological Survey, pp. 89–134.
Stern, R.J., 1981. Petrogenesis and tectonic setting of late Precambrian ensimatic volcanic rocks. Central Eastern Desert. Egypt: Precambrian Res. 16, 168–172.
Stern, R.J., Gottfried, D., Hedge, C.E., 1984. Late Precambrian rifting and crustal evolution in the North Eastern Desert of Egypt. Geolgy 12, 168–172.
Stuckless, J.S., Nkomo, LT., Wenner, D.B., Vantrump, G., 1983. Geochemistry and uranium favourablity of the postorogenic granites of the Northwestern Arabian Shield, Kingdom of Saudi Arabia. Bull. Fac. Earth Sci, King Abdulaziz Univ. 6, 195–209.
Tassinari, C.C.G., Barreto P.M.C., 1992. Uranium in granitoids: recognition criteria of uranium provinces in Brazil. Report No.IAEA-TECDOC-650, 1321.IAEA, Vienna.
UN, 1991. Report on the Seventh IAEA On-Site Inspection in Iraq under Security Council Resolution 687, 14 November 1991, UN Doc S/23215.
Uranium, 1989. OECD/Nuclear Energy Agency and the International Atomic Energy (IAEA): Resources, Production and Demand. OECD/NEA, Vienna, Austria.
Uranium, 1999. OECD/Nuclear Energy Agency and the International Atomic Energy (IAEA): Resources, Production and Demand. OECD 2005/NEA No. (66 2000 07 1 p), p. 340.
Uranium, 2003. OECD/Nuclear Energy Agency and the International Atomic Energy (IAEA), 2003: Resources, Production and Demand. OECD 2005/NEA No.(66 2004 08 1 p), p. 288.
Uranium, 2005. OECD/Nuclear Energy Agency and the International Atomic Energy (IAEA), 2005: Uranium 2005: Resources, Production and Demand. OECD 2005/NEA No. 6098, p. 388.
Ustaömer, T., Robertson, A., 1997. Tectonic-sedimentary evolution of the north Tethyan margin in the Central Pontides of northern Turkey. In: Robinson, A.G. (Ed.), Regional and Petroleum Geology of the Black Sea and Surrounding Region. AAPG Memoir, Tulsa, Oklahoma, 68, pp. 255–290.
Wassef, A.S., 1977. On the results of geological investigations and ore reserves calculations of Abu Tartur phosphorite deposits. Ann. Geol. Surv. Egypt 9, 1–60.
Wilpolt, R.H., Simov, S.D., 1979. Uranium deposits in Africa. In Uranium deposits in Africa: geology and exploration.
Wynn, J., 1996. Geophysics applied to phosphate exploration in northern Saudi Arabia, Presented at the Society of Exploration Geophysicists Annual Meeting, 12 November 1996. Denver, Colorado.
Xoubi, N, 2008. Recent developments of uranium exploration in Jordan (Presentation), Proc. IAEA Technical Meeting on 'Uranium Exploration and Mining Methods', November 17-20, 2008. Amman, Jordan.

Index

Page numbers followed by "*f*" and "*t*" indicate, figures and tables respectively.

F

G

Kolibri 360 GmbH
Ferdinand-Jühlke-Straße 7
99095 Erfurt, Deutschland
produktsicherheit@kolibri360.de